BIOMECHANICS - THEORY AND APPLICATIONS

BIOMECHANICAL VARIABLES IN ASSESSMENT OF FRACTURE RISK

BIOMECHANICS - THEORY AND APPLICATIONS

Additional books in this series can be found on Nova's website under the Series tab.

Additional E-books in this series can be found on Nova's website under the E-book tab.

BIOMECHANICS - THEORY AND APPLICATIONS

BIOMECHANICAL VARIABLES IN ASSESSMENT OF FRACTURE RISK

JAROSLAVA WENDLOVA

Nova Biomedical Books
New York

NOTICE TO THE READER

Library of Congress Cataloging-in-Publication Data

Wendlova, Jaroslava.
Biomechanical variables in assessment of fracture risk / Jaroslava Wendlova.
p. ; cm.
Includes bibliographical references and index.
ISBN 978-1-61470-184-2 (softcover)
1. Fractures--Risk factors. 2. Fractures--Slovakia--Epidemiology. 3. Osteoporosis--Complications--Risk factors. 4. Human mechanics. 5. Women--Diseases--Slovakia. I. Title.
[DNLM: 1. Osteoporotic Fractures. 2. Biomechanics. 3. Risk Assessment. 4. Women's Health. WE 250]
RD101.W46 2011
617.1'5--dc23

2011020240

Published by Nova Science Publishers, Inc. †New York

Contents

Preface

Applied Biomechanics in Bone Densitometry and Musculoskeletal Medicine

Applied biomechanics has recently been achieving an important position within medical disciplines, as it enables a precise quantification of physiological and pathophysiological processes in the organism, expressed in SI units. It introduces the accuracy of technical disciplines into medicine, even though the measurement of physiological and pathophysiological processes in the organism is much more complicated than the measurement of solid, elastic and viscoelastic bodies in technical mechanics.

Experimental, clinical biomechanics and biomedical engineering have been developing quite progressively and it enables a precise measurement of the quality of bones, cartilage, muscles and tendons in vitro and in vivo. This interdisciplinary specialisations obtain an important position in medical disciplines such as osteology, orthopaedics, traumatology, stomatology, rehabilitation - kinesitherapy, orthotics, forensic medicine, in physiology and pathophysiology and step by step they penetrate in other medical disciplines. Nowadays it is possible to measure both linear and non-linear elasticity modules as well as viscoelasticity modules of tissues of the musculoskeletal system with the finite elements method by 3D CT (computer tomography) and 3D micro CT.

These modern experimental methods of bone quality measurement, however, are not so far from their implementation into medical practice!

The objective of my treatise is to bring the importance of measuring the biomechanical variables closer to physicians – osteologists. These biomechanical variables are parts of routine measurement programmes for bone densitometres DXA (dual energy X-ray absorbtiometry), however, they are minimally used in practice. The reason for this phenomenon is probably no full understanding of the importance of biomechanics in osteology by the medical public. And what is not understood, it is often also not noticed.

The monograph presents the results of epidemiological studies realized upon extensive samples from the Slovak population of women in six chapters. The clinical epidemiological studies are original and present the first data into the world literature:

- on expected frequency of values of geometric variables of proximal femur of Slovak women with osteopenia and osteoporosis,
- on expected frequency of biomechanically adverse values of geometric variables of proximal femur in the population of Slovak women,
- on expected frequency of pathological values of the femur strength index (FSI) in the population of Slovak women with osteopenia, osteoporosis and osteoporosis risk factors,
- on the Slovak regression predictive model, enabling to calculate a chance of femoral neck fracture by fall even at those densitometrical workplaces not having the programme for FSI measurement,
- on expected number of women with osteoporosis and osteopenia in the area of proximal femur (total hip) in the population of Bratislava women,
- on estimating the relative risk of femoral neck fracture based on the evaluation of values of T-score plus Z-score variables in the population of Bratislava women with osteoporosis.

The seventh chapter deals with biomechanical variables characterising the viscoelasticity of intervertebral disc, with the elasticity and strength of the vertebra, and with biomechanical analysis of interactive relations of functional motor segments of the spine (vertebra – disc – vertebra), increasing the risk of osteoporotic vertebrae fractures. Interactive pathological relations between vertebrae and discs are classified as:

1. degenerative changes of the intervertebral disc (chondrosis)

- the decrease of the disc viscoelasticity
- the increase of compressive and tensile stress in the disc

2. osteoporosis of vertebrae

- the decrease of vertebrae elasticity and strength
- a phenomenon of local increased compressive stress in the site of vertebra fracture. In the infraction or fracture site of the vertebra body endplate there occurs the locus minoris of the resistance, contributing to the progression of breaking the vertebra and its compression by loading the spine with compressive force.

The contribution of the seventh chapter is the application of this biomechanical analysis into practice in the form of the recommendation – to assess from the X-ray picture of the spine not only the vertebrae deformations in each osteoporotic patient, but also vertebrae endangered by the endplate infraction. The vertebrae at risk include those having between them an intervertebral disc afflicted by chondrosis. In these patients an early and long-term therapeutical intervention is recommended: kinesitherapy under professional guidance and the nutrition therapy.

Quality of osteoporotic bone, bone elasticity and strength – these are the terms physicians are confronted with more and more often in professional osteological literature as well as at scientific congresses, however, for many of them the essence of these terms still remains unclear. In the eighth chapter I explain these terms in detail by biomechanical methods of testing quality of biomaterial, used for determining the bone quality. Without these basic terms the physicians would not be able to understand the importance of biomechanical variables in bone densitometry and in estimating the fracture risk in osteoporotic bone.

The final ninth chapter brings a survey of statistical methods, written in a form easily comprehensible for physicians, for determination of relative risk, absolute risk and the chance of osteoporotic fracture together with calculations for model cases from clinical practice. I deemed it necessary to include this chapter in the book as the problem of clinical osteology is the problem of prevention of bone fractures, and the correct assessment of risk patients with potential spontaneous fracture or fracture by fall is the objective of every clinical osteologist. The second reason for writing this chapter has been the fact that the main criterion of all presented and published prospective interventional clinical studies with antiosteoporotic drugd is the decrease of relative and absolute risk of fracture in the patients. The explanation of terms such as relative risk, absolute risk, difference of risks, odds and odds ratio, makes the comprehension of the results of clinical intervention drug studies closer to physicians.

The book is the result of several years of intensive study and systematic research work. It brings abroad a current view of the problems of osteoporosis in Slovakia.

For Slovak physicians it presents an opportunity of a deeper and wider insight of knowledge in the field of applied biomechanics in musculoskeletal medicine and to show them the importance of biomechanics in clinical medicine. I hope, that applied biomechanics will find more and more interested persons among physicians and it enables the better connecting of biomechanics with clinical osteology and faster development of this interdisciplinary specialisation.

Ass. Prof. Jaroslava Wendlová, MD, PhD.
Author

Reviewer for clinical osteology:
Prof. MUDr. Jaroslav Blahoš, DrSc.

Reviewer for applied biomechanics in bone densitometry:
Doc. RNDr. Štefan Varga, CSc.

All chapters published in this monographs were published in the peer – reviewed scientific journals. Each article (chapter) was reviewed with two independent reviewers.

Professional Profile

Jaroslava Wendlová graduated from the Medical Faculty of Comenius University in Bratislava, Slovakia. After graduation she worked as a lecturer at the Geriatric Clinic and the 1st Internal Clinic in Bratislava. She was awarded the academic degrees of PhD at the Charles Medical University in Prague, Czech Republic, associate professor of internal medicine at Medical Faculty of Slovak University of Health in Bratislava and associate professor of public health at University of Public Health in Trnava, Slovakia.

She was a research fellow at University Clinics in Prague, Vienna, Bad Pyrmont, Hanover, Ulm. She is the first author of 52 original scientific papers and 12 scientific surveys; she is the solo author of 2 scientific monographs, 10 chapters in scientific monographs and 14 professional brochures, her papers have been cited in 83 publications, she presented her papers as the first author at 85 symposia. She is a member of the Slovak Society for Osteoporosis and Metabolic Bone Diseases, as well as the Czech Society for Metabolic Skeletal Diseases.

Based on her publication activities, she was appointed an associate editor of journal Public Health (BioMedCentral, UK), member of the editorial board of Emergency medicine (Dove Medical Press, NZ), Journal für Mineralstoffwechsel (Austria), Wiener Medizinische Wochenschrift (Springer, Austria), active reviewer of WSEAS (World Scientific and Engineering Academy and Society). She was appointed as a Guest editor for a monothematic issue Osteoporosis: Basic and Clinical Aspects (2004) and Applied Biomechanics in Osteology (2010) in Journal Wiener Medizinische Wochenschrift (Springer, Austria).

In the years 2000, 2006 and 2007 the *Society for Metabolic Skeletal Diseases of the Czech Medical Society of J. E. Purkyne* in Prague awarded her the first prize for the best scientific paper in osteology, published in peer – reviewed scientific journals.

In 2005, she successfully completed the international Certified Course in Bone Densitometry of the International Osteoporosis Foundation (IOF) in Vienna and in 2009, the international certified Biomechanical Summer Course of Basic Biomechanics and Biomechanical Methods for Experimental Research of the Musculoskeletal System at Technical University in Ulm, Germany.

In the year 2009 the Presidium of Slovak Medical Association awarded her the Bronze Medal for the development of *a new interdisciplinary scientific branch in Slovakia – biomechanics, applied in clinical osteology.* The management of sanofi – aventis Pharma, Ltd awarded her Certificate of Merit for a series of independent, original, innovative, and successful publications on problems of osteoporosis, contributing to the education of physicians and patients in Slovakia.

She is foundress and president of ongoing Slovak Bioregenerative Medicine Association, which is oriented on bioregeneration of ill organism and support and maintentance health of healthy organism by means of nutrition therapy, applied biomechanics in kinesitherapy and psychohygiene.

Based on her original scientific publication activities, she was nominated to publish her professional profile in the book *Who is Who in the World 2010, 2011? (USA)* and in *Outstanding 2000 Intellectuals of the 21st Century 2010, Publishing House* International Biographical Centre, Cambridge, England.

At present she works in a clinical osteological surgery, lectures superior lectures at the Vienna´s Medical University, University of Public Health and Social Work, Faculty of Physical Education and Sport in Bratislava, acts abroad as a consultant for biomechanical solutions of osteoporotic skeleton, leads seminars and coordinates scientific and research papers in this field.

Chapter I

Slovak Logistic Regression Predictive Model in Estimate of Femoral Neck Fracture by Fall (*Epidemiological Study*)

Abstract

Patients and methods: In the sample of 3, 216 East Slovak women with primary or secondary osteoporosis, osteopenia or risk factors for osteoporosis aged 20 – 89 years, $\widetilde{x} = 58.9$, 95% C.I. (58.42 ; 59.38), we analysed the data obtained from DXA (dual energy X–ray absorptiometry, Prodigy – Primo, GE, USA) measurements of the left proximal femur. The following variables were measured: categorial variable *FSI (femur strength index)* and dependent variables *T-score total hip, BMI (body mass index), alpha angle, theta angle, HAL (hip axis length).* Objectives: To determine from logistic regression the order of dependent variables according to the intensity of their influence (statistical significance) upon the values of the categorial *FSI* variable and to determine how many times raises (declines) the odds of *FSI* acquiring a pathological value (odds for femoral neck fracture by fall), if the value of the variables T – *score total hip, BMI, alpha angle, theta angle and HAL* raises by one unit. Statistical analysis: Logistic regression.

Results: Logistic regression model determined the order of dependent variables according to the intensity of their influence upon the categorial *FSI* variable values: 1. *BMI*, 2. *theta angle*, 3. *T-score total hip*, 4. *alpha angle*, 5. *HAL*. One unit rising of a dependent variable influences with statistical significance the rising (decline) of the odds of categorial *FSI* variable value. Increase by 1^0 of α *angle* variable value increases the odds of *FSI* acquiring a pathological value 1.111 times, increase by 1^0 of θ *angle* variable value increases the odds 1.231 times, increase by 1mm of the *HAL* variable value increases the odds 1.043 times, increase by 1.0 kg/m^2 of the *BMI* variable value increases the odds 1,302 times, increase by + 1 SD of the $T-$ *score total hip* variable value decreases the odds 1,98 times. Conclusion: The Slovak Regression model demonstrates, that the dependent geometric variables of proximal femur statistical significant influence the odds for femoral neck fracture by fall and from logistic regression model we can estimate how many times raise the odds for femoral neck fracture by fall if the geometric variable values raise by one unit.

Keywords: osteoporosis – femoral neck fracture – logistic regression – absolute risk – bone densitometry – femur geometry – biomechanics

Introduction

Bone mineral density (*BMD*), as measured by bone densitometer DXA (dual energy X-ray absorptiometry) has been found to be a reliable index of absolute risk of osteoporotic fractures [1–4]. Prior authors have described the assessment of future fracture risk by combining *BMD* (quantitative variable) values with other relevant patient information (quantitative variables) that has been found to be predictive of fracture [5–7]. More recent models have included the use of logistic regression as a means of identifying the group of risk factors that most strongly predict fracture. A logistic regression model is a type of generalized linear regression model, whereby a group of independent variables can be assembled that best predict a dependent categorical variable (typically fracture in the aforementioned models). The independent (predictive) variables in the model can be quantitative or qualitative. The dependent (predicted) variable is categorical and typically qualitative. In the case of binary logistic regression the dependent variable is dichotomous, with only two possible values, 0 or 1, indicating the absence or presence (respectively) of the outcome of interest. A special logistic regression model can also be created solely from dependent variables [8].

In evaluation of osteoporotic fracture risk, the Femur Strength Index (*FSI*) has been shown to be a reliable index of the odds of fracture. Thus *FSI* values can serve as a validated proxy for fracture risk when assessing the utility of a predictive model [9]. *FSI* computation requires proprietary software used in conjunction with DXA equipment (called Hip Strength Analysis [HSA] software), and incorporates a measure of the geometric distribution of bone in the femoral neck (minimum cross sectional area) called the Cross-Sectional Moment of Inertia (*CMSI*), derived from x-ray absorption curves, along with other measures of bony structural integrity. [10]

In the present paper we describe a methodology for assessing fall-related femur fracture risk called the Slovak Regression Model (SRM). The model differs from previously described fracture models in that it employs quantitative variables of biomechanical and geometric measurements, and results in the prediction of femoral neck fracture by fall, thus providing a tool for the individualized assessment of contemporaneous threat of femur fracture secondary to the most common cause of such fractures.

Objective of the Study

1. To determine from binary logistic regression the ranked order of predictive variables according to the intensity of their influence (statistical significance) on the value of the categorical outcome *FSI* variable.
2. To determine from the model the effect on the odds of *FSI* acquiring a pathological value, defined as *FSI* <1, when the value of the variables *T–score total hip, BMI, alpha angle, theta angle and HAL* increase by a single relevant whole unit (*i.e.* 1 degree, 1 SD, 1 millimetre).
3. To provide clinicians with an alternative means of assessing fall-related femur fracture risk based on easily assessed clinical, demographic and geometric findings without the need for the software required to evaluate *FSI.*

Patients and Methods

A sample of 3,216 women from East Slovakia with primary or secondary osteoporosis, osteopenia and risk factors for osteoporosis aged 20 – 89 years,

$\widetilde{x} = 58.9$, 95% C.I. (58.42 ; 59.38), were examined with the bone densitometer DXA (Prodigy – Primo, GE, USA), CV (coefficient of variation) 1,08%. (Table 1).

Table 1. Characteristic of the sample, followed variables and applied methods

Number	3, 216
Gender	female
Age (median, 95% C.I.)	Median 58.90 95% C.I. (58.42;59.38)
Disease	Osteoporosis, osteopenia, risk factors for osteoporosis
Variables	FSI, age, BMI, angle alpha, angle theta, HAL, T – score total hip
Method	Bone densitometry (DXA, GE, Prodigy Primo, USA)
Statistical analysis	Logistic regression – predictive model

The *BMD* was determined in the standard region of interest (ROI) – total hip left. *BMD* values were given in absolute numbers in *g of Ca-hydroxyapathite crystals for cm*^{2} *(g/cm*^{2}*),* as well as in relative numbers as T-score (the number of standard deviations from the reference group of young healthy women). Osteoporosis or osteopenia were diagnosed in accordance with the WHO criteria:

Classification of bone density	T- score:
Normal	> -1.0 SD
Osteopenia	$\langle -1.0; -2.5)$ SD
Osteoporosis	≤ -2.5 SD

To maintain reliability in the DXA measurement quality (QA) all subjects were scanned on the same device, operated by one of two experienced technicians. The following variables were measured or calculated by the proprietary software: [11]

Theta angle (θ); defined as the angle formed by the femoral neck axis and the femoral shaft axis.

Alpha angle (α); defined as the angle formed by the femoral shaft axis and the perpendicular. The alpha angle can be either positive or negative, depending whether the femur is in a valgus or varus position (Figure 1)

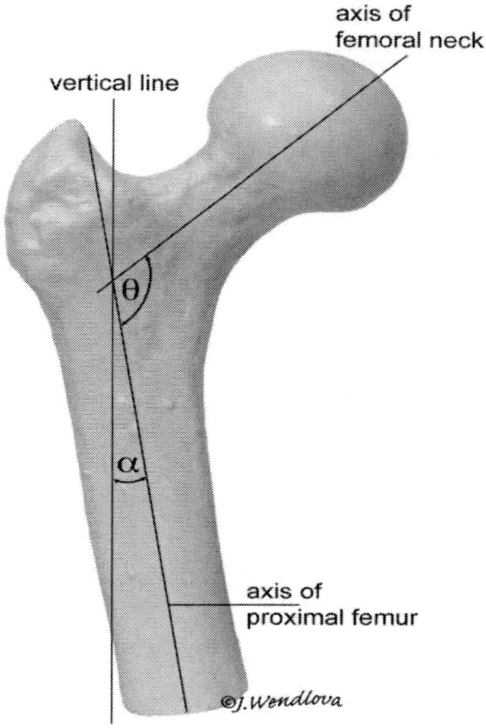

Figure 1. Angle alpha and theta.

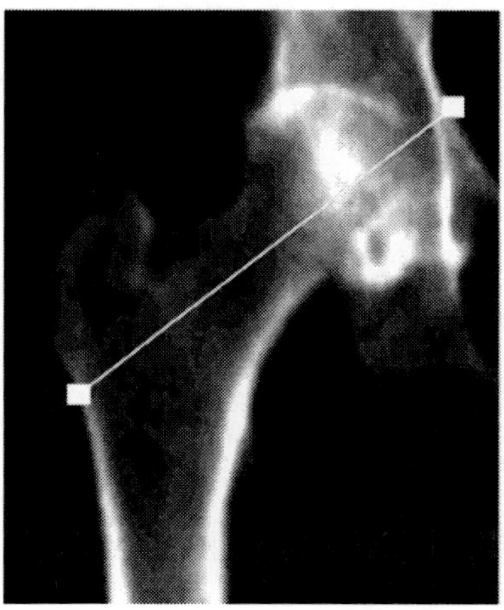

Figure 2. Hip axis length.

HAL; defined as the distance (mm) from the inferolateral aspect of the greater trochanter to the pelvic inner rim, measured along the long axis of the femoral neck (Figure 2)

FSI (femur strength index); defined as the ability or strength of femoral neck to resist compressive (fracture) forces during a fall. The *FSI* is the ratio of estimated elastic limit stress in compression of the femoral neck to the expected compressive stress on the femoral neck caused by a fall onto the greater trochanter, adjusted for the patient´s age, height and weight. The *FSI* integrates four structural characteristics of the femur: *BMD*, bone geometry, bone elasticity, and strength in the femoral minimum neck area with the magnitude of impact force resulting from a predicted fall on the greater trochanter, and taking into account the patient´s height and weight. A normal value of *FSI* is defined as ≥ 1, and a pathological value is < 1.

Minimum cross sectional area (mCSA); defined as the smallest cross sectional area (mm^2) in the femoral neck, also known as the "dangerous cross section area" for its propensity to fracture.

In addition to these variables, body mass index (*BMI*) was calculated for all of the subjects. *BMI* is defined as $\frac{weight(kg)}{height^2(m^2)}$.

Statistical Analysis

The statistical analysis used was logistic regression, utilizing commercially available statistical analysis software (SAS® *Enterprise Guide 4.0*). [8, 12-14].

Logistic Regression

The results of a clinical study, including 3,216 East Slovak female patients, were used to set up a model. To analyze the data of the sample the statistical methods were applied using statistical programme systems SAS® *Enterprise Guide 4.0*.

The Slovak Binary Regression Model

- *Categorial (dichotomous) variable* in the model was the *FSI* variable, whose values were categorized as follows: *FSI* > *1* (normal values);

in the regression model these values were replaced by the 0 category, categorized *FSI* ($FSI_{categor} = 0$) [8,12-14].
FSI < 1 (pathological values); in the regression model these values were replaced by the 1 category, categorized *FSI* ($FSI_{categor} = 1$). A binary variable was produced with the working title $FSI_{categor}$ in the outputs from *SAS Enterprise Guide.* [8,12-14]
The sample included 470 (14.51%) patients with pathological values of *FSI* ($FSI_{categor} = 1$) and 2,746 (85.37%) patients, whose *FSI* values could be considered as normal ($FSI_{categor} = 0$) (Table 2).

- *Dependent variables*, shortlisted for the regression model, included: *BMI, age, alpha angle, theta angle, HAL, T-score total hip.*
  The method of stepwise regression was applied in the selection of the most optimal subset of dependent variables for regression model. The dependent variable age was excluded from the list of chosen dependent variables .

Table 2. Frequency of occurrence of categorized FSI variable ($FSI_{category}$) in the sample

Response Profile			
Ordered Value	$FSI_{categor}$	Total Frequency	In %
1	1	470	14.61
2	0	2746	85.37

Table 3. Testing of model's statistical significance

Testing Global Null Hypothesis: β = 0			
Test	Chi-Square	DF	p
Likelihood Ratio	658.1840	5	<.0001*
Score	564.5674	5	<.0001*
Wald	431.7499	5	<.0001*

p-probability.
DF-degree of freedom.
* statistical significance.

Testing the Statistical Significance of the Model [8]

The model with the variables *BMI, alpha angle, theta angle, HAL, T-score total hip* is statistically significant as a whole. At the significance level of α = 0.01 and on the basis of the likelihood ratio test, test score and Wald test, we can reject the zero hypothesis H_0: the model as a whole is not statistically significant (all regression coefficients equal zero), and accept an alternative hypothesis H_A: the model as a whole is statistically significant (at least one regression coefficient does not equal zero) (Table 3).

Verification of Model Quality

The ability of the model to differentiate between the patients with normal and those with pathological values of *FSI*, based on the values of dependent variables, was verified by the Hosmer – Lemeshow test. At the significance level of $\alpha = 0.01$ we accepted the zero hypothesis H_0: the model is adequate to the data. This was tested against an alternative hypothesis H_A: the model is not adequate to the data (Table 4) [8].

Table 4. Testing model's quality

Hosmer and Lemeshow Goodness-of-Fit Test		
Chi-Square	DF	p
12.6364	8	0.1250

$\alpha = 0.01$, $p > \alpha$ ($0.1250 > 0.01$).

Results

The Table 5 contains dependent variables arranged according to the intensity of their influence upon the occurrence of values of a categorial *FSI* variable, evaluated by the *Wald' s* test of statistical significance. All selected independent variables affect the categorial *FSI* variable with statistical significance already at the level of $\alpha = 0.001$.

The regression model equation allows a probability estimate of the occurrence of pathological values of *FSI* $[P\ (FSI_{categor} = 1)]$ in the femoral neck area, i. e., the probability estimate of femoral neck fractures, caused by fall (the relation is valid for the population of Slovak women).

Estimated values of *OR (odds ratio)* are given in the Table 6. None of the estimated 95% confidence intervals (*CI*) contains the value of 1. Therefore, it holds true for each dependent variable that one unit change of a given variable affects with statistical significance the change of the odds (increase or reduction of the odds) that the *FSI* categorial variable acquires pathological values. The growth of dependent variable values *BMI, alpha angle, theta angle, and HAL* increases the odds. If the value of the *BMI* dependent variable increases by one unit (+1.0 kg/m^2) and the values of remaining dependent variables do not change, the odds that categorial variable $FSI_{categor}$ acquires in the regression model the value of 1 (i.e., pathological value of *FSI* < *1*) increases 1.302 times. In the same way it is possible to interpret the estimated values of variables *alpha angle, theta angle, and HAL.*

The increase of values in the dependent variable *T-score total hip* leads to the reduction of the odds that the *FSI* categorial variable acquires pathological values. If the value of the dependent variable *T-score total hip* increases by one unit (+ 1 SD) and the values of remaining dependent variables do not change, the odds that categorial variable $FSI_{categor}$ acquires the value of 1 (i. e., pathological value of *FSI* < 1) declines to the level 0.505 of the original values, i. e., declines about 1.98 times.

Table 5. Estimated model's variables and testing of their statistical significance

Analysis of Maximum Likelihood Estimates					
Variable	DF	Estimate β	Standard Error	Wald Chi-Square	p
Intercept	1	-40.7612	2.4387	279.3576	<.0001*
BMI	1	0.2639	0.0148	316.6411	<.0001*
angle theta left	1	0.2082	0.0173	145.0853	<.0001*
T - score total hip left	1	-0.6824	0.0603	127.9164	<.0001*
angle alpha left	1	0.1051	0.0177	35.0879	<.0001*
HAL left	1	0.0423	0.00959	19.4942	<.0001*

Table 6. Regression model allows to interpret the values of dependent variables through OR (odds ratio) values

Odds Ratio Estimates			
Variable	Point Estimate	95% Wald Confidence Limits	
T - score total hip left	0.505	0.449	0.569
angle alpha left	1.111	1.073	1.150
angle theta left	1.231	1.190	1.274
HAL left	1.043	1.024	1.063
BMI	1.302	1.265	1.340

Discussion

In the past few years several papers were published about regression models created with the aim to determine a probability estimate (absolute risk) for osteoporotic fractures with the prognosis for five, ten or more years [15 - 17]. However, their authors admit that these regression models have their inaccuracies and shortcomings. The exposure to some risk factors is not precisely defined, as for example: smoking, alcohol consumption, secondary diseases negatively affecting the bone metabolism (increased osteoresorption), glucocorticoid usage, etc. There is no possibility to include into the specification of the regression model, within these qualitative risk factors, the length of exposure to a risk factor and in some the size of its dose according to an individual data from the patient. Another disadvantage of regression models is the fact that they were created without involving the population of the age ≥ 80 years, even though it is known that just this group most often suffers from femoral neck fractures as well as other fractures. The risk estimate of fractures with the prognosis for five or ten years is not the most appropriate one, as most fractures happen by fall and not spontaneously. The regression models do not take into consideration whether the osteporotic patient (male or female) exercises regularly according to the instructions of a rehabilitator:

- exercises for balancing the muscular dysbalance [18]
- exercises with simulated disruption of balance, with the aim to improve the coordination movements
- exercises with the correct techniques of falling (distribution of the energy of fall into several consequent phases of the movement)

The risk of a fall for an osteoporotic patient, who does not exercise regularly the abovementioned exercises and prefers a sedentary lifestyle, is higher than for a patient with a regular physical activity according to the instructions of a rehabilitator.

How to Estimate Correctly the Probability (Absolute Risk) of a Femoral Neck Fracture (Figure 3, Table 7)

Previously used regression models have been based on various qualitative risk factors and only on one characteristics of the bone quality – *BMD*. Criteria for some risk factors cannot be defined so precisely as to meet fully the physician's requirement for real individual characteristics of the tested patient, and so the new regression models get entangled in adding more and more new risk factors, belonging primarily to qualitative variables. The inaccuracy of qualitative risk factors definitions is joined by incorrect estimate of the spontaneous fracture risk with the prognosis for different future time intervals and not with the prognosis of fracture risk by fall. In the estimated future interval the intensity of risk factors can change and it is also not taken into consideration. We based the Slovak regression model on an assumption that all risk factors, without direct necessity of their knowledge, listing, and characteristics, share in the formation of the resulting bone quality and the bone quality determines the absolute risk of fracture by fall. Therefore, in the Slovak regression model we focused on as accurate determination as possible of material properties of bones (considering our diagnostic possibilities) – their strength and elasticity. The bone quality was determined by biomechanical quantitative variables. Falls usually happen by a certain stereotypical mechanism, allowing biomechanics to calculate accurately the magnitude of impact force by fall upon the hip [18]. This impact force magnitude is taken into account in the biomechanical FSI variable. Therefore, in the Slovak regression model an estimate is made for the fracture probability by fall. It can happen that after getting the prognosis of absolute risk of a spontaneous fracture, say, within 5 years, on the next day the patient can fall and suffer a fracture of a vertebra or a femoral neck.

The structure of the Slovak regression model and its solution represents a new approach to the probability (absolute risk) estimate for femoral neck fractures by fall. In the Slovak regression model, *BMI* is the statistically most significant dependent variable affecting the categorial *FSI* variable. It means that *BMI* significantly influences the probability of femoral neck fracture by fall. The explanation is simple. We put the values of variables for height and weight into the equation for the calculation of impact force by fall and this equation is part of the equation for *FSI* calculation [18]. It holds true that the higher the weight or height of the patient, the bigger the impact force applied to the hip, increasing the probability of fracture by fall. If the patient has higher values of variables alpha angle, theta angle, *HAL* than the mean value (μ) in the Slovak population, her probability of femoral neck fracture by fall is also higher.

The age was not decisive in our regression model; therefore, the programme excluded it from the model. For the regression model it was sufficient to give a current value of *T- score total hip (BMD)* and *FSI* value, characterising the bone quality.

Table 7. Two approaches how to prognosticate the femoral neck fracture. Estimate of absolute risk (probability) in the future and accurate forecast by fall at presence

Logistic regression predictive models: questionnaire	Bone densitometry: biomechanical variables
Estimate of absolute fracture risk (AFR) based on: 1. clinical risk factors only (RF) 2. RF + *BMD* .	Forecast of fracture origin based on: 1. *FSI* values (assessment of bone quality -- biomechanical variables of proximal femur)
Estimate of AFR in the future with indicated time interval. The mode of fracture is not defined. By fall? Spontaneous fracture?	Forecast of fracture origin by fall at present.
Intensity and speed of exposure to RF is not constant. The intensity and speed change of exposure to RF, applied in time, is not taken in account when estimating the fracture risk	The forecast always considers the current status of quality of bone and its present ability to resist impact force by fall
AFR can every year change in dependence of intensity change and speed change in exposure to RF, therefore it is wrongly to do the estimate for the future (5 – 10 years forwards) coming out of stationary RF .	Forecast is always actual by each bone quality measurement
1. The annual estimate of AFR is the same, if the number of RF is constant and the regression model is based only on RF. 2. If the regression model is based on RF + BMD and the number of RF is constant, the annual estimate of AFR depends only on BMD value changes.	The annual forecast of fracture origin by fall depends on bone quality changes (on biomechanical variable values)
1. Estimate based on *only qualitative dichotomous variables* (classified as "yes" or "no") – model based on RF only 2. Estimate based on *qualitative dichotomous variables* and one *quantitative variable BMD* measured in SI units – model based on FR + BMD	Forecast based on *quantitative variables only, measured in SI units.*
Are we interested in an uncertain estimate of AFR in future, while we cannot define the mode of fracture ? By fall? Spontaneous fracture?	Are we interested in a current accurate forecast of fracture origin by fall at present?

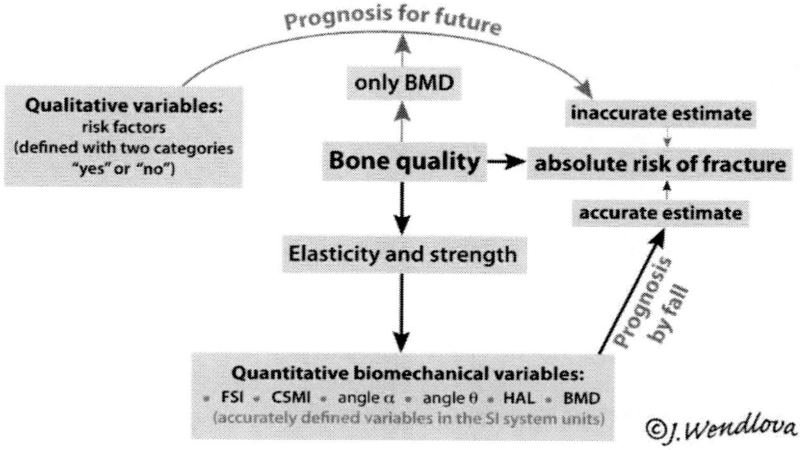

Figure 3. Two Possibilities How to Estimate the Probability (Absolute Risk) of Femoral Neck Fracture.

BMD and *FSI* values are, however, age-dependent, but their value can be at any age affected by exposure to a number of risk factors; therefore, even at a younger age it can acquire very low pathological values, e.g., depending on the length of exposure of the patient to diseases pathologically increasing osteoresorption (e. g., thyreotoxicosis, chronical hepatopathy, chronical pancreatitis,etc.). Geometric variables of proximal femur *(*alpha angle, theta angle, HAL*) + BMD (T-score total hip) + BMI* variable enable us to estimate the probability of the occurrence of pathological values of *FSI* in the femoral neck area.

At present, *FSI* is the only available variable, corresponding to biomechanical criteria of loading the bone by fall, whose values can be determined by DXA in a routine ambulatory praxis. *FSI* integrates in itself three important bone characteristics:

1. BMD
2. bone geometry: alpha angle, theta angle, HAL, CSMI (cross-sectional moment of inertia in the minimum cross section area of femoral neck)
3. elasticity and strength

Whether the *FSI* variable value is calculated from the two-dimensional bone measurement by DXA, which in comparison with three-dimensional measurement has some inaccuracies, it is a value determining the bone quality much more precisely than *BMD*. Clinical importance of the *FSI* variable value lies in the fact that it enables to determine an accurate individual risk for femoral neck fractures by fall.

The values of the categorial *FSI* variable, lower than 1, increase the probability of a fracture in the femoral neck area by fall [19 – 23]. The lower the value of the *FSI* variable than 1, the bigger the probability for a fracture by fall. The equation of the Slovak regression model makes it possible in praxis to determine the probability for a acquiring the pathological value of *FSI* at those densitometrical workplaces, not having a programme for measuring the *FSI* variable. The doctor puts into the regression model equation (given in Excel® programme) the current values of the patient: *BMI, angle alpha, angle theta, HAL, standardized BMD (sBMD) measured in the total hip area.*

Standardized BMD (s BMD) total hip:

- sBMD = 1 000 x k x BMD
- k – coefficient given by GE

CONCLUSION

We have opened the way for a new method of estimating the probability or the odds for a femoral neck fracture by fall, based upon bone quality determination.

The Slovak regression model:

- demonstrates, that the dependent geometric variables of proximal femur statistical significant influence the odds for femoral neck fracture by fall and from logistic regression model we can estimate how many times raise the odds for femoral neck fracture by fall, if the geometric variable values raise by one unit,
- estimates the present quality status of femoral neck by means of variable value *FSI,*
- estimates the probability of femoral neck fracture by fall at present on the basis of the values of *quantitative* biomechanical variables of proximal femur, which are given in the units of the International System of Units (SI),
- eliminates the inaccuracies of other models, based on not entirely precisely defined *qualitative* variables plus BMD, which are not able

to determine the probability of femoral neck fracture by fall at present,

- makes it possible in praxis to determine the probability (absolute risk) of femoral neck fracture by fall also at those densitometrical workplaces, not having a programme for measuring the *FSI* variable.

The study investigated East Slovak women. Although it can be supposed that similar results will be found in the whole Slovak female population, in the other female populations at least in central Europe. Further studies will have to verify this.

Estimate of FSI in the practice can be helpful in the strategy of health economics as it may determine the patients with increased risk of femoral neck fractures by fall. For these risk patients it is necessary to apply an intensive complex therapeutical intervention (drug therapy, kinesitherapy, nutritional therapy, wearing a hip protector) which proves to be effective in the prevention of costs reimbursement for the therapy of femoral neck fractures.

Declaration – Conflict of Interests

The author declares that she has no competing interests (financial or non financial).

References

[1] Melton LJ III. Wahner HW. Richelson LA. O'Fallon WM. et al: Osteoporosis and the risk of hip fracture. *Am. J. Epidemiol.* 1986; 129(5): 254-61.

[2] Kanis JA. Johnell O. Oden A. Jonsson B. Dawson A. Dere W: Risk of hip fracture derived from relative risk: An analysis applied to the population of Sweden. *Osteoporosis Int.* 2000; 11(2): 120–7.

[3] Kanis JA. Mc Closkey EV. Johansson H. et al: Case finding for the management of osteoporosis with FRAX assessment and intervention thresholds for the UK. *Osteoporos Int.* 2008;19(10):1395-408.

[4] Kanis JA. Johnell O. Oden A. Oglesby AK. De Laet C. Jonsson B: Ten year probabilities of osteoporotic fracture according to BMD and diagnostic thresholds. *J. Bone Miner Res.* 2001; 16: (Suppl 1), 156-7.

[5] Kanis JA: Diagnosis of osteoporosis and assessment of fracture risk. *Lancet* 2002; 359(9321): 1929–36.

[6] Black DM. Arden Palermol NK. Pearson J. Cummings SR.: Prevalent vertebral deformities predict hip fractures and new vertebral deformities but not wrist fracture: Study of osteoporotic fractures research group. *J. Bone Miner Res.* 1999; 14(5): 821–8.

[7] Klotzbuecher CM. Ross PD. Lansman PB. Abbot TA. Berger M: Patients with prior fracture have an increased risk of future fractures: a summary of the literature and statistical synthesis. *J. Bone Miner Res.* 2000; 15(4) : 721–39.

[8] Kleinbaum DG, Klein M: Statistics for biology and health. Logistic Regression. 2nd ed, Springer *Science USA* , 2002. pp. 1-124.

[9] Faulkner KG, Wacker WK, Barden HS, Simonelli C, Burke PK, Ragi S, Del Rio L. Femur strength index predicts hip fracture independent of bone density and hip axis length. *Osteoporos Int.* 2006;17(4):593-9.

[10] Yoshikawa T, Turner CH, Peacock M, Slemenda CW, Weaver CM, Teegarden D, Markwardt P, Burr DB. Geometric structure of the femoral neck measured using dual-energy x-ray absorptiometry. J Bone Miner Res. 1994 Jul;9(7):1053-64. Erratum in: *J. Bone Miner Res.* 1995 Mar;10(3):510.

[11] Nakamura T. Turner CH. Yoshikawa T. Slemenda CW. et al: Do variations in hip geometry explain differences in hip fracture risk between Japanese and white Americans? *J. Bone Miner Res.* 1994; 9(7): 1071-76.

[12] Varga S: Robust Estimations in Classical Regression Models Versus Robust Estimations in Fuzzy Regression Models. *Kybernetika* 2007; 43(4): 503-8.

[13] Hosmer DW. Lemeshow A.: *Applied logistic regression.* Publishing house: John Wiley and Sons, 2 nd ed., 120–135, 2000, ISBN: 0471356328

[14] Varga S: Another View on the Fuzzy Regression. *Forum Statisticum Slovacum* 2009; 10(3): 1–7.

[15] Kanis JA. Johnell O. Oden A. Johansson H. McCloskey E.: FRAXTM and the assessment of fracture probability in men and women from UK. *Osteoporos Int.* 2008; 19(4): 385–97.

[16] Fujiwara S. Nakamura T. Orimo H. Hosoi T. Gorai I. et al: Development and application of a Japanese model of the WHO fracture risk assessment tool (FRAXTM) *Osteoporos Int.* 2008; 19(4): 429–35.

[17] Siris E. Delmas D: Assessment of 10 – year absolute fracture risk: a new paradigm with worldwide application. *Osteoporos Int.* 2008; 19(4): 383–4.

[18] Wendlová J: Why is so Important to Balance the Muscular Dysbalance in mm. coxae Area in Osteoporotic Patients? *Bratisl. lek. listy* 2008; 109 (11): 502-7.

Chapter II

Expected Frequency of Femoral Neck Fractures by Fall in the Osteoporotic and Osteopenic East Slovak Female Population *(Epidemiological Study)*

Abstract

Objective of the Study: To compare an expected frequency of pathological *FSI* values with an expected frequency of pathological BMD (*T-score* ≤ - 2.5 SD) in the total hip area in the East Slovak female population.Patients and Methods: We analysed the data in the sample of 3,215 East Slovak women with a primary or secondary osteopenia and osteoporosis and with risk factors for osteoporosis, aged 20 –89 years, $\widetilde{x}$ = 58.9 years, 95% C. I. (55.15; 60.35). Measured variables: *T – score total hip left, FSI left (femur strength index* with dual energy X-ray absorptiometry. Results: In the East Slovak female population we can expect 6, 25% of women with osteoporosis in the total hip area and 14.54% of women with *FSI* < *1*, i.e. these women will suffer a femoral neck fracture in case of a fall. The expected frequency of the incidence of pathological *FSI* values of the East Slovak female population is 2.33 times higher than the expected frequency of the incidence of pathological *T - score* for osteoporosis measured in the total hip area. Conclusion: The measurement of *FSI* variable values may discover a higher percentage of women with a risk of femoral neck fracture by fall than the measurement of *BMD* variable values in the total hip area. This disproportion shows that even a patient with osteopenia measured in the total hip area may sustain a femoral neck fracture by fall, when she has adverse values of geometric variables of proximal femur (biomechanically unfavourable proximal femur configuration).

Keywords: osteoporosis - femur strength index (FSI) - cross sectional area (CSA) – cross sectional moment of inertia (CSMI) – femoral neck fracture – dual energy X-ray absorptiometry (DXA)

INTRODUCTION

Spheres of interest, objectives and directions of epidemiological studies have been changing with the development of clinical osteology. At first, the interest focused mainly upon determining the percentage of osteopenia and osteoporosis incidence in the male and female populations in different countries. Comparisons were made of ethnic, geographic and nutritional influences on osteoporosis incidence in different countries and the reasons of incidence differences were investigated.

Rising costs of complex treatment of fractures and social care of patients with permanent consequences of fractures (reduced work skills, reduced self-sufficiency, disability) brought to attention 1) the precise assessment of bone quality and 2) the prevention strategy of osteoporotic fractures [1, 2]. Preparation of background data for economic strategy in the osteoporosis management required clinical studies determining optimal *BMD* (bone mineral density) values at which the medicinal treatment is effective in the fracture prevention, as well as selecting medications enabling effective reduction of fracture risk [3]. Results of these clinical studies made it possible to develop a strategy when to initiate an anti-osteoporotic therapy and which medications to apply. The treatment costs of fractures caused by falls represent still an economic burden for the health care system [4 – 6]. Therefore, the search continues to find a more precise method to determine patients with a fracture risk by fall. Recently, regression models started to be used in the clinical practice for this purpose [7, 8]. They can determine an absolute risk (probability) of fracture for a patient, including the prognosis for the future.

The mode of fracture origin is not defined (by fall? spontaneous?). The regression models take into consideration only risk factors for the development of osteoporosis or risk factors plus the *BMD* value in the total hip area. The downside of the now used regression models is that they cannot estimate the probability of femoral neck fracture by fall, but only of a spontaneous fracture with the prognosis for the future for different time intervals. This paper is focused upon an analysis of expected frequency of *FSI* (femur strength index) variable values in the *East Slovak female population of osteopenic, osteoporotic women and women with risk factors for osteoporosis (hereafter East Slovak female population). FSI* is a biomechanical variable determining whether the bone strength in the femoral neck area endures the load of compressive force impact by fall (normal value: $FSI \geq 1$, pathological value: $FSI < 1$).

OBJECTIVES OF THE STUDY

1. To estimate an expected frequency of the occurrence of pathological *FSI* variable values in the East Slovak female population and so to assess an expected frequency of femoral neck fractures by fall in these women.
2. To estimate an expected frequency of the occurrence of pathological *BMD (T–score ≤ - 2.5 SD)* variable values in total hip area in the East Slovak female population.
3. To compare an expected frequency of the occurrence of pathological *FSI* variable values with an expected frequency of osteoporosis *(T-score)* occurrence measured in the total hip area in the East Slovak female population.

PATIENTS AND METHODS

Characteristics of the Sample

Using a DXA bone densitometer (Prodigy – Primo, GE, USA) we analysed the data in the sample of 3,215 East Slovak women aged 20 – 89 years, $\widetilde{x} = 58.9$ years, 95% C. I. (55.15; 60.35):

1. with risk factors for the development of osteoporosis
2. with a primary or secondary osteopenia
3. with a primary or secondary osteoporosis

who were examined in the years 2007 – 2009:

Characteristics of Variables and Applied Methods

To achieve the measurement quality (QA), only two operators alternated in measuring with the DXA device and all women were measured with the same device. The following biomechanical variables were measured in the area of the left proximal femur: *BMD total hip (T-score total hip), CSA (cross-sectional area); CSMI (cross-sectional moment of inertia), and FSI (femur strength index).*

All women were examined with the same bone densitometer DXA. The *BMD* was determined in the standard region of interest (ROI) – total hip left. *BMD* values were given in absolute numbers in *g of Ca-hydroxyapathite crystals for* cm^2 (g/cm^2), as well as in relative numbers as T-score (the number of standard deviations from the reference group of young healthy women). Osteoporosis or osteopenia were diagnosed in accordance with the WHO criteria.

Characteristics of the Population

All women of the East Slovakia with osteopenia, osteoporosis and with risk factors for osteoporosis (hereafter East Slovak female population).

Definitions of Measured Variables

CSA (cross sectional area) – is the minimum cross sectional area (in mm^2) in the femoral neck area (dangerous cross section).

CSMI (cross sectional moment of inertia) – is defined as the sum of multiplications of elementary areas and the squares of their distances from neutral axis, denoted as I_Y (given in cm^4). The larger the cross sectional area, the higher the number of small elementary areas (A) and so is the bigger the second root of the distance (z^2) of these elementary areas from the neutral axis. The enlargement of the cross sectional area is accompanied by the enlargement of the cross sectional moment of inertia (equation 1).

$$I_Y = \int_{A_1}^{A_n} z^2 \cdot dA(cm^4) \quad (1)$$

A_{1-n} – elementary area A_1 to A_n

z – distance of relevant elementary area from neutral axis

FSI (femur strength index) – characterizes the strength of femoral neck, stressed by compressive forces during a fall. It is defined as a ratio of estimated elastic limit in compression of the femoral neck (σ_E) to the expected compressive stress of a fall on the greater trochanter adjusted for the patient´s age, height and weight (σ_C) [9].

STATISTICAL ANALYSIS

To analyse the data of the sample, statistical methods were applied using statistical programme systems Statgraphics Centurion XV.

1. We tested the character of the distribution of *CSA, CSMI* and *FSI* variable values in the sample and determined the incidence of the frequency of measured variable values in 10-year age intervals.
2. Using the Goodness-of-Fit Test χ^2 (chi-Squared Test) and coming from the character of empirical distribution of the frequency (probability) of *FSI* variable values in the *sample*, we estimated the expected frequency of *FSI* variable values in *the East Slovak female population* at the significance level $\alpha = 0.01$ [10 – 11].
3. We tested the character of the distribution of variable values in the sample and using the Goodness-of-Fit Test χ^2 (chi-Squared Test) and coming from the character of empirical distribution of the frequency (probability) of *T-score total hip* variable values in the *sample,* we estimated the expected frequency of *T-score total hip* variable values in the *East Slovak female population* at the significance level $\alpha = 0.05$ [12].
4. To illustrate the results we used the following statistical graphs [13]:

- Histograms
- Quartile – Box and Whisker Plots

RESULTS

Age Variable

The median $(\widetilde{x})$ for age is 58. 9 years, 95% C.I. for the median (55.15; 60.35). One quarter of patients from the sample are in age < 57.7 years and one quarter of patients are in age > 67.2 years.

In the sample, the patients were divided in 10 – year age intervals. In each interval (age group) there were evaluated quartiles, medians, averages of variables *CSA, CSMI, FSI.*

CSA Variable

The Figure 1 (Box and Whisker Plot) shows the statistically significant decrease of the average $(\overline{x})$, median $(\widetilde{x})$, the second (Q_2^4) and third (Q_3^4) quartile of the *CSA* variable values, associated with increasing age.

While the median $(\widetilde{x})$ value for *CSA* in the age group 20 to 30 years is 140.529 mm², in the age group above 80 years it reaches the value of 106.404 mm² only.

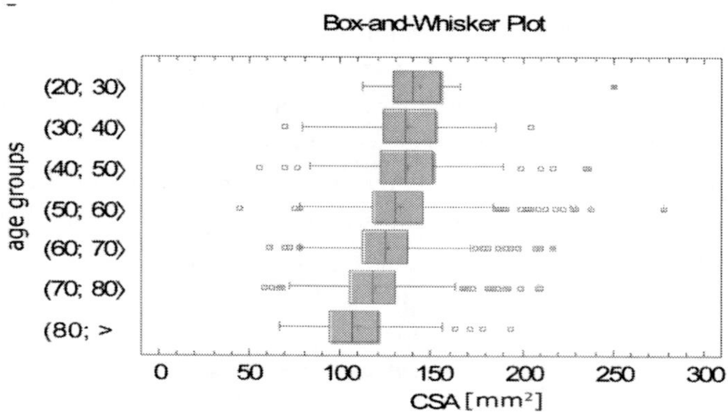

Figure 1. Box and whisker plots for CSA in age groups.

CSMI Variable

The Figure 2 (Box and Whisker Plot) shows decrease of the average $(\bar{x})$, median $(\tilde{x})$, the second (Q_2^4) and third (Q_3^4) quartile of the variable values, associated with increasing age. This decrease is not as significant as by *CSA*, because the CSMI reflects not only the cross sectional area, but also the geometry of the cross section area.

While the median $(\tilde{x})$ value for *CSMI* in the age group 20 to 30 years is 8, 929.17 mm^4, in the age group above 80 years it reaches the value of 7, 952.99 mm^4 only.

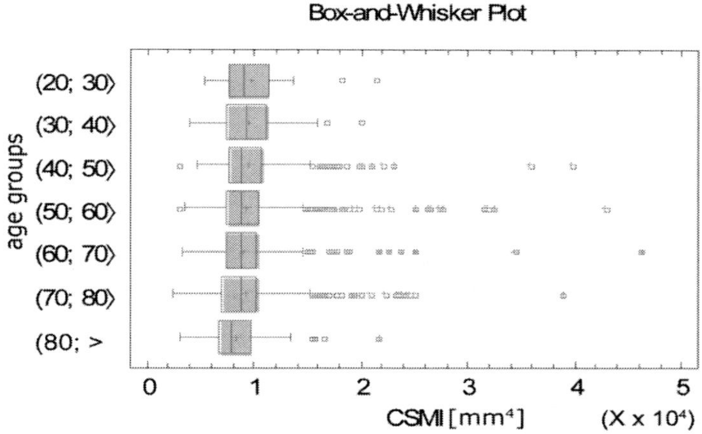

Figure 2. Box and whisker plots for CSMI in age groups.

FSI Variable

The Figure 3 shows decline of the average $(\bar{x})$, median $(\tilde{x})$, the second (Q_2^4) and third (Q_3^4) quartile of the variable values, associated with increasing age. This decrease is not as significant as by *CSA* or *CSMI,* firstly in the age groups 60 – 70, 70 – 80 and above 80 years, in which medians and averages do not change. Following the analysis of these age groups we have discovered, that in these age groups, there were more patients with biomechanically favourable configuration of the proximal femur. Therefore the values of variable *FSI* were advantageous.

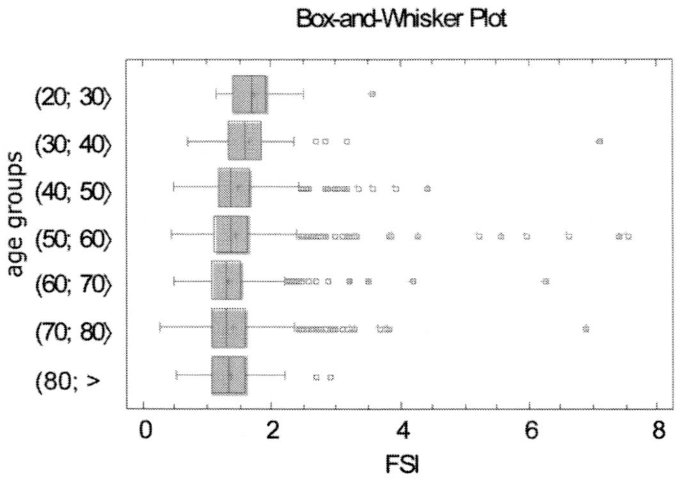

Figure 3. Box and whisker plots for FSI in age groups.

While the median $(\widetilde{x})$ value for *FSI* in the age group 20 to 30 years is 1.715, in the age group above 80 years it reaches the value of 1.351 only.

Table 1. Observed and expected frequencies for *FSI* in East Slovak female population. (Chi-Squared Test, Goodness-of-Fit Test)

	FSI					
	Lower Limit	*Upper Limit*	*Observed Frequency*	*Expected Frequency*	*Expected Freq. in %*	*Chi-Squared*
at or below		0.5	6	5.77	0.18	0,01
	0.5	1.0	465	461.44	14.36	0,03
	1.0	1.5	1614	1644.68	51.17	0.57
	1.5	2.0	859	812.14	25.27	2.70
	2.0	2.5	191	202.48	6.30	0.65
	2.5	3.0	45	55.46	1.73	1.97
	3.0	3.5	17	18.37	0.57	0.10
	3.5	4.0	6	7.12	0.22	0.18
	4.0	4.5	3	3.12	0.10	0.00
above	4.5		8	3.42	0.11	6.13

Chi-Squared = 12. 3499 with 6 d.f. P-Value = 0.0546021.
Goodness-of-Fit Tests, Chi-Squared Test.
Sample: n = 3, 214 (from the sample n = 3, 215 one error value FSI = 0 was excluded).

The Table 1 brings an empirical distribution of *FSI* variable values frequencies in the sample and the expected distribution of the *FSI* values frequencies in the East Slovak female population. In the East Slovak female population we can expect 0.18% of women with the *FSI* values < 0.5 and 14.36% of women with the *FSI* values from the interval (0.5; 1.0). It means that 14.36% + 0.18% = 14.54% of women from the East Slovak female population will sustain a femoral neck fracture by fall. At the significance level $\alpha = 0,01$ it can be assumed that the probability distribution for *FSI* variable values in the East Slovak female population is loglogistic with the parameters derived on the basis of values found in the sample (Figure 4). The hatched field in the Figure 4 represents the expected number of frequencies of *FSI* values < 1.

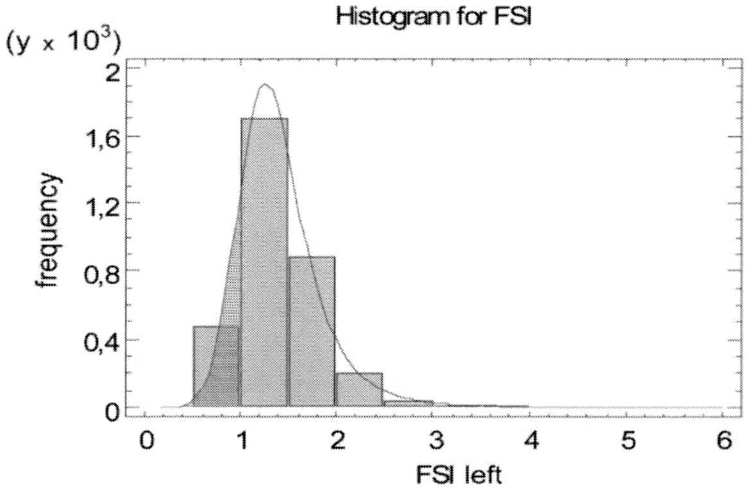

Figure 4. Expected distribution of frequencies for FSI variable values in the East Slovak population of osteopenic , osteoporotic women and women with risk factors for osteoporosis. The hatched field represents an expected number of frequencies for FSI < 1.

T-Score Total Hip Variable

In the East Slovak female population according to values measured in the total hip area we can expect 6.25% of osteoporotic women with the T-score values ≤ 2.5 SD and 34.94% of osteopenic women with T-score values from the interval (- 2.5; -1.0) SD. At the significance level α = 0,05 it can be assumed that the probability distribution for *T-score total hip* variable values in the East Slovak female population is loglogistic with the parameters derived on the basis of values found in the sample (Table 2, Figure 5).

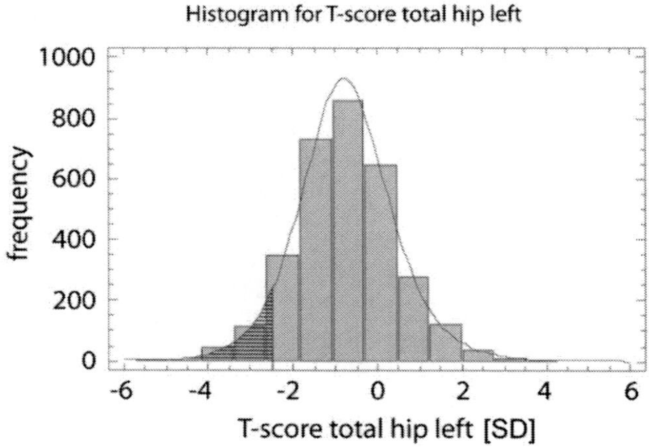

Figure 5. Expected distribution of frequencies for T – score total hip variable values in the East Slovak population of osteopenic , osteoporotic women and women with risk factors. The hatched field represents an expected number of frequencies for T – score ≤ - 2.5 SD.

Table 2. Observed and expected frequencies of *T – score total hip* in East Slovak female population

T–score total hip left (SD)	Observed frequency in the sample	Expected frequency in the sample in %	Expected frequency In the population in %
(- ∞; - 2.5⟩	196	6.10	6.25
(- 2.5; - 1.0⟩	1142	35.52	34.94
(- 1.0; 0.0⟩	1089	33.87	34.59
(0.0;+ ∞)	788	24.51	24.22

Sample: n = 3,215.

Discussion

The study proves a statistically significant reduction of *CSA* and *CSMI* of femoral neck with aging. This reduction is not so significant for *FSI,* because the value of *FSI* is dependent also on the values of geometrical variables of proximal femur, which can be biomechanically favourable or biomechanically unfavourable. The values of geometrical variables of proximal femur are born with and not dependent on aging.

At present, *FSI* is the only available variable, corresponding to biomechanical criteria of loading the bone by fall, whose values can be determined by DXA in a routine ambulatory practice. *FSI* integrates in itself three important bone characteristics:

1. BMD
2. bone geometry
3. elasticity and strength

Whether the *FSI* variable value is calculated from the two-dimensional bone measurement by DXA, which in comparison with three-dimensional measurement has some inaccuracies, it is a value determining the bone quality much more precisely than *BMD.* Clinical importance of the *FSI* variable value lies in the fact that it enables to determine an individual risk for femoral neck fractures by fall. The prediction of femoral neck fractures by fall on the basis of *BMD* is markedly limited, as *BMD* characterises only a small part of bone quality and is a only one of a number of risk factors for femoral neck fractures.

Authors of some clinical retrospective and prospective studies [14 – 20] agree that the increase of the values of geometric variables of proximal femur • *HAL (hip axis length)*, • *theta angle* --defined as an angle formed by the femoral neck axis and the femoral shaft axis, • *d* (average width of femoral neck) over the mean value in the population is a risk factor for femoral neck fractures.

Faulkner et al. [21] compared *BMD, HAL, CSMI* and *FSI* obtained from DXA measurement in a group of women with and without hip fracture. Femoral neck *BMD* and *FSI* were significantly lower and *HAL* significantly higher in fractured group compared with controls. Mean *CSMI* was not significantly different between fracture patients and controls. After adjustment of these variables in regression model, the authors conclude that *BMD, HAL* and *FSI* are significant independent predictors of hip fracture.

The objective of our clinical study was to determine an expected frequency of the *FSI* pathological values incidence in the population of East Slovak female population. We were interested in the expected percentage of osteopenic and osteoporotic women who would sustain femoral neck fractures by fall. The asset of our study is the finding that the expected percentage of *FSI* pathological values incidence (14.54%) is 2.33 times higher than the expected percentage of osteoporosis incidence in the total hip area (T-score ≤ -2.5 SD) in the East Slovak female population (Table 3).

Table 3. Comparison of expected frequencies of pathological values of variables *FSI* with *T – score total hip* in East Slovak female population

Variable	Observed frequency in sample	Expected frequency in population (%)
FSI left < 1 n = 3 214	471	**14,54**
T–score ≤ - 2,5 SD (total hip left) n = 3215	196	**6,25**

It is evident even from such a simple comparison that *BMD* and *FSI* variables are two different characteristics of bone quality. *FSI,* which includes in itself the *BMD* variable, finds a higher percentage of women at risk for femoral neck fracture by fall than *BMD*. This disproportion shows that even a patient with osteopenia measured in the femoral neck area may sustain a femoral neck fracture by fall, in case she has adverse values of geometric variables of the proximal femur *(biomechanically unfavourable proximal femur configuration).*

The *FSI* value in clinical practice enables not only to determine an individual risk for femoral neck fracture, but helps to improve the strategy of treatment in the prevention of femoral neck fractures.

The *FSI* value < 1 could therefore be used as a criterion for the initiation of preventive therapeutic interventions in order to avoid femoral neck fractures. These interventions may include:

1. drug therapy
2. wearing a hip joint protector
3. kinesitherapy:
    - coordination exercises
    - exercises to remove the muscular dysbalance in the mm. coxae area [22]
    - training of techniques to fall correctly.

Methods of measurement and determination of bone quality have been constantly improving. As demonstrated by the latest clinical studies, the future of bone densitometry lies in a three-dimensional measurement of *FSI* and other new progressive biomechanical variables [23 – 27].

Conclusion

New types of DXA and QCT densitometers provide programs for clinical and ambulatory practice enabling to measure different biomechanical and geometric varibales of proximal femur. Regrettably, these programs are utilized minimally in practice and osteologists are still rigidly adhering to *BMD* values. Offered programs overrun the "guidelines" of international and national osteological societies, which still have not included the criteria for the evaluation of new biomechanical variables.

Summing Up

1. *FSI* is a biomechanical variable determining whether the bone strength in the femoral neck area endures the load of compressive force impact by fall. It expresses the bone quality much more accurately than *BMD*.
2. The measurement of *FSI* variable values may discover a higher percentage of women in the population with a risk of femoral neck fracture by fall than the simple measurement of *BMD* variable values in the total hip area.
3. Patient with osteopenia measured in the total hip area may sustain a femoral neck fracture by fall, when she has adverse values of

geometric variables of proximal femur (biomechanically unfavourable proximal femur configuration).

4. Pathological *FSI* value should become a criterion in the osteological praxis for the onset of complex treatment of patients in the prevention of femoral neck fractures.
5. The study investigated East Slovak women. Although it can be supposed that similar results will be found in the whole Slovak female population and in other female populations at least in central Europe, further studies will have to verify this.

Declaration – Conflict Of Interests

The author declares, that she has no competing interests (financial or non financial).

References

[1] Pietschmann P, Kerschan – Schindl K : Knochenqualität – wissenschaftliche Aspekte versus praktische Relevanz. *J. Miner Stoffwechs*. 2004;11:16–8.

[2] Boonen S, Singer AJ: Osteoporosis management: impact of fracture type on cost and quality of life in patient at risk of fracture. *Curr. Med. Res. Opin*. 2008;24:1783 – 90.

[3] Kanis JA, Oden A, Johnell O et al: The burden of osteoporotic fractures: a method for setting intervention thresholds. *Osteoporos. Int.* 2001;12:417–27.

[4] Finnern HW, Sykes DP: The hospital cost of vertebral fractures in the EU: estimates using national data sets. *Osteoporos. Int.* 2003;14:429–36.

[5] Lindsay R, Burge RT, Strauss DM: One year outcomes and costs following a vertebral fracture. *Osteoporos. Int.* 2005;16:78—85.

[6] Rabenda V, Manette C, Lemmens R et al.: The direct and indirect costs of he chronic management of osteoporosis: a prospective follow- up of 3440 active subjects. *Osteoporos. Int.* 2006;17:1346 – 52.

[7] Kanis JA, Johnell O, Oden A et al: FRAXTM and the assessment of fracture probability in men and women from UK. *Osteoporos. Int.* 2008;19:385–97.

[8] Fujiwara S, Nakamura T, Orimo H, Hosoi T, Gorai I et al: Development and application of a Japanese model of the WHO fracture risk assessment tool (FRAXTM). *Osteoporos. Int.* 2008;19:429–35.

[9] Nakamura T, Turner CH, Yoshikawa T et al: Do variations in hip geometry explain differences in hip fracture risk between Japanese and white Americans? *J. Bone Miner Res.* 1994;9:1071–6.

[10] Pacáková V: *Statistical methods for economists*. 2nd ed. Bratislava: Iura Edition; 2009. pp. 174-177.

[11] Varga S: Another View on the Fuzzy Regression. *Forum Statisticum Slovacum* 2009; 3:1–7.

[12] Kleinbaum DG, Klein M: *Statistics for biology and health*. Logistic Regression. 2nd ed, Springer Science USA , 2002. pp. 1-124.

[13] Rublíková E, Labudová V, Sandtnerová S: *Analysis of categorical data.* (Analýza kategoriálnych údajov), 1st ed. Bratislava: Economic University Bratislava; 2009. pp 41 – 141.

[14] Kukla C, Gaebler C, Pichl RW et al.: Predictive geometric factors in a standardized model of femoral neck fracture. Experimental study of cadaveric human femurs. Injury, *Int. Care Injured* 2002;33:427–33.

[15] Gregory JS, Testil D, Stewart A et al.: A method for assessment of the shape of the proximal femur and its relationship to osteoporotic hip fracture. *Osteoporos. Int.* 2004;15:5–11.

[16] Alonso Gomez C, Curiel Diaz M, Carranza Hawkins F et al.: Femoral bone mineral density, neck shaft angle and mean femoral neck with as predictors of hip fracture in men and women. *Osteoporos. Int.* 2000;11:714–20.

[17] El Kaissi S, Pasco JA, Henry MJ et al.: Femoral neck geometry and hip fracture risk: the Geelong osteoporosis study. *Osteoporos. Int.* 2005;16:1299– 303.

[18] Watts NB: Fundamentals and pitfalls of bone densitometry using dual – energy X – ray absorbtiometry (DXA). *Osteoporos. Int.* 2004;15:847–54.

[19] Gnudi S, Malavolta N, Testi D, Viceconti M: Differences in proximal femur geometry distinguish vertebral from femoral neck fractures in osteoporotic women. *The Brit. J. Radiol.* 2004;77:219–23.

[20] Gnudi S, Ripamonti C, Lisi L et al.: Proximal femur geometry to detect and distinguish femoral neck fractures from trochanteric fractures in postmenopausal women. *Osteoporos. Int.* 2002;13:69–73.

[21] Faulkner KG, Wacker WK, Barden HS et al.: Femur strength index predicts hip fracture independent of bone density and hip axis length. *Osteoporos. Int.* 2006; 17:593–9.

[22] Wendlová J: Why is so Important to Balance the Muscular Dysbalance in mm. coxae Area in Osteoporotic Patients? *Bratisl. lek. listy* 2008;109:502–7.

[23] Cheng X, Li J, Lu Y et al: Proximal femoral density and geometry measurements by quantitative computed tomography. Association with hip fracture. *Bone* 2007;40:169– 74.

[24] Manske SL, Liu Ambrose T, De Bakker PM et al: Femoral neck cortical geometry measured with magnetic resonance imaging is associated with proximal femur strength. *Osteoporos. Int.* 2006;17:1539– 45.

[25] Bousson V, Le Brass A, Roqueplan F et al: Volumetric quantitative computed tomography of the proximal femur: relationships linking geometric and densitometric variables to bone strength. Role for compact bone. *Osteoporos. Int.* 2006;17:855– 64.

[26] Engelke K, Adams JE, Armbrecht G et al: Clinical Use of Quantitative Computed Tomography and Peripheral Quantitative Computed Tomography in the Management of Osteoporosis in adults. The 2007 ICSD Official Position. *J. Clin. Densit.* 2008;11:123-62.

[27] Hansma P, Yu H, Schultz D, Rodriguez A et al: The tissue diagnostic instrument. *Rev. Sci. Instrum.* 2009;80(5):34-38.

Chapter III

Expected Frequency of Biomechanically Adverse Values of Proximal Femur Geometric Variables in the East Slovak Female Population (*Epidemiological Study*)

Abstract

Patients and Methods: DXA (dual energy X – ray absorbtiometry) measurements of the left proximal femur were obtained and analysed from sample 3,151 East Slovak women aged 20 – 89 years, $\widetilde{x}$ = 58.9 years, 95% C. I. (55.15; 60.35) with risk factors for the development of osteoporosis, with osteopenia and osteoporosis using bone densitometer (Prodigy – Primo, GE, USA). Measured variables: 1. geometric variables: *alpha angle, theta angle, HAL (hip axis length)* and 2. biomechanical variable: *FNS (femoral neck strength).*

Objective: 1. In the female population to determine the expected frequency of biomechanically adverse values of variables *alpha angle, theta angle, and HAL*, increasing the risk of a femoral neck fracture by fall, 2. To determine from the logistic regression how many times are the odds of femoral neck fracture by fall higher, if the value of variables *alpha angle, theta angle, and HAL* raises by one unit.

Statistical analysis: 1. Goodness-of-Fit Test χ^2 --test (chi-Squared Test). 2. Logistic regression.

Results: Biomechanically adverse value of *angle* $\alpha > 6.869^o$ can be expected in 10% of the East Slovak female population; extremely adverse value of angle $\alpha > 12.3^o$ in 1%; biomechanically adverse value of *angle* $\theta > 129.405^o$ in 20%; extremely adverse value of *angle* $\theta > 132.290^o$ in 10%; biomechanically adverse length of *HAL > 109.930 mm* in 20%; extremely adverse length of *HAL > 113.015 mm* in 10%.

Conclusion: 1. According to the data in our study, the physicians are able to distinguish the *biomechanically adverse* and *extreme adverse* values of proximal femur geometric variables for femoral neck fracture by fall measured by DXA in the patients from Slovak female population 2. These patients should be treated, except the drug therapy, with special care: wearing a hip protector, kinesitherapy to remove the muscular dysbalance in the mm. coxae area, training of techniques how to fall correctly.

Keywords: osteoporosis – geometric variables – proximal femur –bone densitometry- epidemiological study – femoral neck fractures – biomechanics

INTRODUCTION

The costs of treatment of fractures caused by falls still represent an economic burden for the health care system. Rising costs of complex treatment of fractures and social care of patients with permanent effects of fractures (reduced work skills, reduced self-sufficiency, and disability) stimulated an interest in the prevention strategy of osteoporotic fractures [1-6]. Therefore, the search continues to find a more precise method of pinpointing patients with a fracture risk by fall.

This paper is focused upon an analysis of expected frequency distribution of proximal femur geometric variable values in the East Slovak female population. As no professional or scientific literature brings average values of geometric variables such as *angle* α and *angle* θ for gender and different races, our objective was to determine these values in the East Slovak female population. The knowledge of point (mean values) and interval (95% CI – confidence interval) estimates for proximal femur geometric variables in the Slovak female population enables physicians to determine female patients with biomechanically adverse configuration of proximal femur geometric variables, which are over the upper limit of 95% C.I. It enables to determine osteoporotic female patients at high risk of a femoral neck fracture by fall in osteological practice.

Objective of the Study

1. To find out *point* (mean values) and *interval estimates* (95% C.I.) of geometric variable values of *alpha angle, theta angle, and HAL* (hip axis length) in the East Slovak female population.
2. To determine the character of *expected frequency distribution of variable values* of *alpha angle, theta angle, and HAL* in the East Slovak female population.
3. To find out the *expected frequency of biomechanically adverse values* of *alpha angle, theta angle, and HAL* variables in the East Slovak female population, elevating the risk of femoral neck fractures by fall.
4. To determine from logistic regression, how many times raises the odds of a femoral neck fracture by fall, if the value of the dependent variables *alpha angle, theta angle and HAL* increases by one whole unit of relevant measurement (i.e. one degree, one millimetre etc.).

Patients and Methods

Characteristics of the Sample

DXA measurements data of the left proximal femur were obtained and analysed in the sample of 3,151 East Slovak women aged 20 – 89 years, $\widetilde{x}$ = 58.9 years, SD = 11,1 without case history of proximal femoral fracture

1. with risk factors for the development of osteoporosis
2. with a primary or secondary osteopenia
3. with a primary or secondary osteoporosis

using bone densitometer (Prodigy – Primo, GE, USA), total CV (coefficient of variation) 1,08%. Variables were determined using the manufacturer's AHA (advanced hip assessment) program that is commercially available. The AHA program automatically calculates a number of bone geometry variables and a biomechanical variable *femur strength index (FSI)* from the scan image and bone distribution variables derived from information contained within DXA X – ray absorption curves [7]. The more precise term for *FSI* is *femoral neck strength (FNS)*, because the stress is calculated in the minimum cross sectional area of neck.

The following variables of the left proximal femur were measured by the DXA device:

1. geometric variables: α angle, θ angle, HAL (hip axis length)
2. biomechanical variable: FNS (femoral neck strength).

Characteristics of the Population

All women [95% C. I. for age (55.15; 60.35)] of the East Slovakia with osteopenia, osteoporosis and with risk factors for osteoporosis without case history of proximal femoral fracture (hereafter *East Slovak female population*).

To maintain reliability in the DXA measurement quality (QA) all subjects were scanned on the same device, operated by one of two experienced technicians.

Definitions of the variables measured or calculated by the proprietary software [8]:

- *Alpha* defined as the angle formed by the femoral shaft axis and the perpendicular. The alpha angle can be either positive or negative, depending whether the femur is in a valgus or varus position (Figure 1)
- *Theta* angle (θ) is an angle formed by the femoral neck axis and the femoral shaft axis (Figure 1).
- *HAL (hip axis length)* – defined as the distance (mm) from the inferolateral aspect of the greater trochanter to the pelvic inner rim, measured along the long axis of the femoral neck (Figure 2).
- *FNS* (femoral neck strength) - defined as the ability or strength of femoral neck to resist compressive (fracture) forces during a fall. The *FSI* is the ratio of estimated elastic limit stress in compression of the femoral neck to the expected compressive stress on the femoral neck caused by a fall onto the greater trochanter, adjusted for the patient´s age, height and weight. The *FNS* integrates four structural characteristics of the femur: *BMD*, bone geometry, bone elasticity,

and strength in the femoral minimum neck area with the magnitude of impact force resulting from a predicted fall on the greater trochanter, and taking into account the patient's height and weight. A normal value of *FNS* is defined as ≥ 1, and a pathological value is < 1.

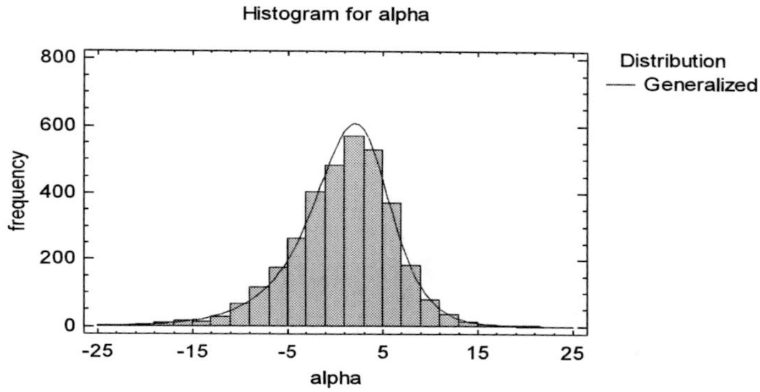

Figure 1. Expected distribution of frequencies for angle alpha variable values is logistic (East Slovak female population).

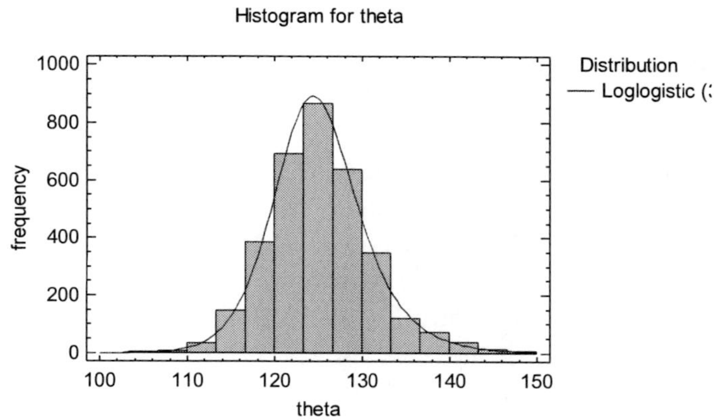

Figure 2. Expected distribution of frequencies for angle theta variable values is loglogistic. (East Slovak female population).

STATISTICAL ANALYSIS [9 –11]

Statistical methods were applied to analyse the data of the sample of East Slovak women using statistical programme systems:

A. Statgraphics Centurion XV.
B. SAS® Enterprise Guide 4.0. (logistic regression)

1. We determined the point (median) and interval estimates (95% C. I.) for the values of variables *α angle, θ angle, and HAL* in the East Slovak female population.
2. Using the Goodness-of-Fit Test χ^2 (chi-Squared Test) and coming from the character of empirical distribution of the frequency (probability) of variable values of *α angle, θ angle, and HAL* in the *sample*, we estimated the expected distribution of frequencies of these variable values in the *population* of East Slovak women at the significance level α = 0.05.
3. On the basis of generalized logistic distribution we determined the expected frequency of biomechanically adverse values of geometric variables of *α angle, θ angle, and HAL* in the East Slovak female population, which raise the risk for developing femoral neck fractures by fall.
4. Applying logistic regression, we determined the odds for femoral neck fracture by fall, if the value of the *α angle , θ angle, and HAL* variables increases by one fhole unit.

RESULTS

The Table 1 brings the point and 95% interval estimates of α *angle, θ angle, and HAL* variable values in the East Slovak female *population.*

It can be expected with the probability of 0.95 (95%). that the mean value (μ) will be in the East Slovak female population:

- for *α angle* from the interval (0.932;1.291)
- for *θ angle* from the interval (124.929; 125.323)
- for *HAL* from the interval (104.642;105.102)

Using the Goodness-of-Fit Test χ^2 (chi-Squared Test) we estimated the expected distribution of frequencies of *α angle, θ angle, and HAL* variable values in the *population* of East Slovak women at the significance level α = 0.05. Expected distribution of variable values frequency for the variable *α angle* in the East Slovak female population is logistic (Graph 1, Table 2), for *θ angle* loglogistic (Graph 2, Table 3), and for *HAL* logistic (Graph 3, Table 4).

Table 1a. Summary statistics for alpha, theta and HAL in the sample

Parameter	alpha (α)	theta (θ)	HAL
Count	3151	3151	3151
Average	1.112	125.129	104.872
Median	1.605	124.747	104.757
Standard deviation	5.139	5.727	6.593
Coeff. of variation	4. 623%	4.577%	6.287%
Minimum	-28.454	105.791	57.914
Maximum	17.892	149.974	130.185
Lower quartile	-1.884	121.31	100.585
Upper quartile	4.472	128.295	108.883
Skewness	-0.636	0.476	-0.135
Kurtosis	1.490	0.913	2.262

Table 1b. Means and 95% confidence intervals for alpha, theta and HAL in the East Slovak female population

Variable	Mean	Stnd. error	Lower limit	Upper limit
alpha	1.112	0.092	0.932	1.291
theta	125.129	0.102	124.929	125.329
HAL	104.872	0.117	104.642	105.102

Table 2. Goodness-of-Fit Tests for alpha Chi-Squared Test

	Lower Limit	*Upper Limit*	*Observed Frequency*	*Expected Frequency*	*Chi-Squared*
at or below		-25.0	1	2.75	1.11
	-25.0	-21.0	3	4.74	0.64
	-21.0	-19.0	4	4.87	0.16
	-19.0	-17.0	9	8.04	0.12
	-17.0	-15.0	16	13.26	0.56
	-15.0	-13.0	14	21.88	2.84
	-13.0	-11.0	26	36.09	2.82
	-11.0	-9.0	65	59.42	0.52
	-9.0	-7.0	113	97.49	2.47

Table 2. (Continued)

	Lower Limit	*Upper Limit*	*Observed Frequency*	*Expected Frequency*	*Chi-Squared*
	-7.0	-5.0	175	158.58	1.70
	-5.0	-3.0	265	252.67	0.60
	-3.0	-1,0	402	383.89	0.85
	-1.0	1.0	481	526.99	4.01
	1.0	3.0	570	600.43	1.54
	3.0	5.0	527	522.36	0.04
	5.0	7.0	371	340.81	2.67
	7.0	9.0	181	177.88	0.05
	9.0	11.0	77	81.36	0.23
	11.0	13.0	35	34.87	0.00
	13.0	15.0	12	14.52	0.44
	15.0	17.0	4	5.97	0.65
above	17.0		2	4.13	1.10

Chi-Squared = 25.139 with 18 d.f. P-Value = 0.121135.

Table 3. Goodness-of-Fit Tests for theta Chi-Squared Test

	Lower Limit	*Upper Limit*	*Observed Frequency*	*Expected Frequency*	*Chi-Squared*
at or below		106.667	1	2.23	0.68
	106.667	110.00	6	9.27	1.16
	110.0	113.333	34	38.38	0.50
	113.333	116.667	145	133.56	0.98
	116.667	120.00	385	371.30	0.51
	120.00	123.333	691	725.34	1.63
	123.333	126.667	865	863.87	0.00
	126.667	130.0	636	625.45	0.18
	130.0	133.333	348	324.48	1.71
	133.333	136.667	119	145.28	4.75
	136.667	140.00	70	62.74	0.84
	140.00	143.333	35	27.41	2.10
	143.333	146.667	14	12.33	0.23
	146.667	150.00	4	5.73	0.52
above	150.00		0	5.64	5.64

Chi-Squared = 21.4173 with 11 d.f. P-Value = 0.029297.

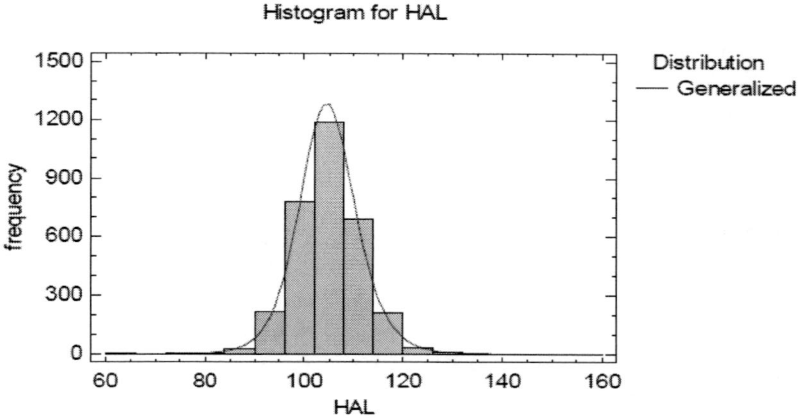

Figure 3. Expected distribution of frequencies for HAL (mm) variable values is logistic (East Slovak female population).

Generalized logistic distribution enabled us to determine the expected frequency (expressed in %) of biomechanically adverse variable values of proximal femur geometric variables in the East Slovak female population. Biomechanically adverse values of proximal femur geometric variables (α , θ , *HAL*) are values higher than the upper limit of 95% C. I. for geometric variables in the East Slovak female population.

Table 4. Goodness-of-Fit Tests for HAL Chi-Squared Test

	Lower Limit	*Upper Limit*	*Observed Frequency*	*Expected Frequency*	*Chi-Squared*
at or below		84.00	6	7.33	0.24
	84.00	90.00	22	36.98	6.07
	90.00	96.00	211	201.72	0.43
	96.00	102.00	781	761.62	0.49
	102.00	108.00	1188	1216.21	0.65
	108.00	114.00	690	677.94	0.21
	114.00	120,00	210	193.57	1.39
	120.00	126.00	35	43.78	1.76
	126.00	132.00	8	9.35	0.20
above	132.00		0	2.50	2.50

Chi-Squared = 13.9466 with 6 d.f. P-Value = 0.0302388.

Angle α (Table 5)

If lower tail area is α ≤ 0. 9, it means that, we can expect 90% of East Slovak female population with angle $\alpha \leq 6.871^0$ and 10% of East Slovak female population with biomechanically *adverse* magnitude of angle $\alpha > 6.871^0$. (See 95% C.I. for angle α).

If lower tail area is α ≤ 0.99 , it means that we can expect 99% of East Slovak female population with angle $\alpha \leq 12{,}3\ ^0$ and 1 % of the East Slovak female population with biomechanically *extremely adverse* magnitude of angle $\alpha > 12.3^0$

Table 5. Generalized logistic distribution for various values of *α* and correspondent values of variable *angle alpha*

Lower Tail Area α ≤	Generalized Logistic angle alpha
0.01	-15.015
0.1	-5.790
0.5	1.358
0.9	6.871
0.99	12.303

Angle θ (Table 6)

If lower tail area is α ≤ 0.80 --- in 20 % of the East Slovak female population we can expect a biomechanically *adverse* magnitude of angle $\theta > 129.405^0$

Table 6. Generalized loglogistic distribution for various values of *α* and correspondent values of variable *angle theta*

Lower Tail Area α ≤	Loglogistic (3-Parameter) angle theta
0.55	125.463
0.6	126.122
0.7	127.574
0.8	129.405
0.9	132.29

If lower tail area is $\alpha \leq 0.90$ — in 10 % of the East Slovak female population we can expect a biomechanically *extremely adverse* magnitude of angle $\theta > 132.29^0$.

Hal (Table 7)

If lower tail area is $\alpha \leq 0.80$ — in 20 % of the East Slovak female population we can expect a biomechanically *adverse* magnitude of *HAL* length > *109.93 mm*

If lower tail area $\alpha \leq 0.90$ – in 10 % of the East Slovak female population we can expect a biomechanically *extremely* adverse magnitude of *HAL* length > *113.015 mm*

From the set of equations [8], of which is the variable *FNS* calculated, we have mathematically deduced, that the stress (σ) in minimum cross sectional area of femoral neck, stressed by compressive force during a fall, is higher when:

- the *angle theta* variable rises,
- the *angle alpha* variable rises,
- *HAL* variable is longer,
- the patient´s *weight* and *height* rise

It means, that the rising of these variable values increase the risk for fracture in femoral neck area by fall. The same results we have obtained from predictive logistic regression model.

Table 7. Generalized logistic distribution for various values of *α* and correspondent values of variable *HAL*

Lower Tail Area α ≤	Generalized Logistic HAL (mm)
0.55	105.485
0.6	106.245
0.7	107.896
0.8	109.93
0.9	113.015

Predictive Logistic Regression Model, Odds Ratio (OR) (Table 8)

The model of a binary (categorical) logistic regression is used to predict the probability of an alteration of a categorical variable *FNS,* conditioned by the values of selected geometric dependent (predictive) variables *α angle, θ angle, HAL.* The logistic regression model as a predictive model allows assessing the intensity of the influence of selected dependent (predictive) variables upon the occurrence of values (categories) of categorical variable *FNS.*

Table 8. Regression model allows to interpret the values of dependent variables angle alpha, theta, HAL through OR (odds ratio) values in relation to categorical FSI variable values

Odds Ratio Estimates			
Dependent Variables	Point Estimate	95% Wald Confidence Intervals	
angle alpha_left	1.111	1.073	1.150
angle theta left	1.231	1.190	1.274
HAL	1.043	1.024	1.063

This table is an abstraction from complete table of regression model published in article from Wendlova J. (13).

The Slovak regression model was dealt with in the article by J. Wendlova and therefore we describe it in brief [12].

The Structure of the Slovak Regression Model

Categorical (dichotomous, binary) variable in the model was the *FNS* variable, whose values were categorized as follows: *FNS > 1* (normal values); in the regression model these values were replaced by the 0 category, categorized *FNS* ($FNS_{categor} = 0$) [10].

FNS < 1 (pathological values); in the regression model these values were replaced by the 1 category, categorized *FNS* ($FNS_{categor} = 1$). The categorical variable *FNS* was created with the working title $FNS_{categor}$ in the outputs from *SAS Enterprise Guide.* [10] *Selected dependent variables*: *α angle, θ angle, HAL.*

Odds Ratio (OR)

Estimated OR values are given in the Table 8. None of estimated 95% C. I. includes the value of 1. Therefore for each dependent variable (*α angle, θ angle, HAL)* it holds true that one unit rise of a given variable influences with statistical significance the odds (increase or reduction of the odds) for categorical *FNS* variable to acquire a pathological value.

If the α *angle* value rises by 1^0 , and the values of remaining variables in the regression model do not change, the odds of *FNS* acquiring a pathological value (the odds of a femoral neck fracture by fall) increases 1.111 times.

If the θ *angle* value rises by 1^0, and the values of remaining variables in the regression model do not change, the odds of *FNS* acquiring a pathological value (the odds of a femoral neck fracture by fall) increases 1.231 times.

If the *HAL* length increases by 1 mm, and the values of remaining variables in the regression model do not change, the odds of *FNS* acquiring a pathological value (the odds of a femoral neck fracture by fall) increases 1.043 times.

DISCUSSION

Several scientists measured independently geometric variables as *HAL, θ, α angle* in clinical studies and they concluded, that the geometric variation of proximal femur significantly contributes to the large variation in hip fracture risk [13 -25]:

1. geometric variables play important roles in estimating fracture risk and should be assessed along with *BMD* when using DXA [14],
2. geometric variables predict hip fracture and their evaluation may be useful in the understanding of the biomechanics of the hip fracture [15, 16],
3. women with neck fracture had a significantly longer *HAL*, lover *CSMI* (cross sectional moment of inertia in minimum cross section of neck) and higher BR (buckling ratio) than controls without fractures [17],
4. *θ angle* can be used to determine of the fracture risk independently of *BMD* [18],
5. patient with osteopenia measured in the total hip area may sustain a femoral neck fracture by fall, when she has adverse values of

geometric variables of proximal femur (biomechanically unfavourable proximal femur configuration) [25],

6. there is a direct relationship between the predictor variable values *BMI (body mass index), alpha angle, theta angle,* and *HAL* and an categorical variable *FNS* value of <1, in that an increase of one relevant unit of measurement in these variables significantly increases the odds of fall-related fracture [12].

The contribution of our study lies in the finding of the 95% C.I., within we can expect with 95% probability the mean values (μ) of proximal femur geometric variables *HAL, θ angle, and α angle* in the East Slovak female population. These data for *angle α* and θ have not been published until now in domestic as well as international literature and have been missing in clinical osteological practice. Now the physicians are able to determine the biomechanically adverse values of proximal femur geometric variables measured with DXA that could represent a risk of a femoral neck fracture by fall. Also, it is important the finding of expected percentage of women with biomechanically adverse or extremely biomechanically adverse values of the variables *α angle, θ angle, and HAL* in the East Slovak female population. *What do we understand under biomechanically adverse values of proximal femur geometric variables?* Biomechanically adverse values of proximal femur geometric variables are values higher than the upper limit of 95% C.I. for measured geometric variable. While for *α angle* we can expect biomechanically extremely adverse values in 1% of the East Slovak female population, for each *θ angle* and *HAL* we can expect 10%. If some of those women with risk factors or osteopenia will develop osteoporosis, they will be at high risk of a femoral neck fracture by fall.

As the geometric variables, including α *angle, θ angle, and HAL,* are given by an anatomical shape of proximal femur, genetically conditioned in each individual, osteoporosis cannot influence these parameters. Therefore, we consider that similar distribution of frequencies for femoral geometric variable values can be expected in the whole female Slovak population.

Although it can be supposed that similar 95% C.I. for geometric variable values of proximal femur will be found in other populations at least in central Europe, further studies will have to verify this.

CONCLUSION

According to the data in our study, the physicians are able to distinguish the *biomechanically adverse* and *extreme adverse* values of proximal femur geometric variables for femoral neck fracture by fall measured by DXA in the patients from Slovak female population. Proximal femur geometric variables should be also measured in patients with osteoporosis and osteopenia along with *BMD* when using DXA. The osteoporotic patients at high risk of femoral neck fractures by fall could be diagnosed according to the finding of biomechanically adverse configuration of the proximal femur. These patients should be treated complex, except the drug therapy, with special care: wearing a hip protector, kinesitherapy to remove the muscular dysbalance in the mm. coxae area and the training of techniques how to fall correctly. Removal of muscular dysbalance in the mm. coxae area is an effective and economic measure within the prevention of complicated comminutive fractures of proximal femur [26].

With our results we wanted to promote and accelerate the introduction of DXA measuring proximal femur geometric variables in outdoor osteological practice and to explain to physicians the importance of these measurements, allowing a more accurate individual estimate of a femoral neck fracture risk by fall. Unfortunately, despite the fact that many types of DXA bone densitometers enable the measurement of geometric and biomechanical variables, their application in practice is minimal.

The biomechanical variable *FNS* is a step forwards from a simple variable *BMD*. But the future is open for three – dimensional proximal femur imaging measurements with finite element method analysis, which reveal the true potential how to identify more precise at – risk individuals for hip fractures [27].

DECLARATION – CONFLICT OF INTERESTS

The author declares, that she has no competing interests (financial or non financial).

REFERENCES

[1] Boonen S, Singer AJ. Osteoporosis management: impact of fracture type on cost and quality of life in patient at risk of fracture. *Curr. Med. Res. Opin.* 2008;24:1783–90.

[2] Stein VK, Dorner TH, Lawrence K, et al. Economic concepts for measuring the costs of illness of osteoporosis: An international comparison. *Wien. Med. Wochenschr.* 2009;159:253-61.

[3] Finnern HW, Sykes DP. The hospital cost of vertebral fractures in the EU: estimates using national data sets. *Osteoporos. Int.* 2003;14:429–36.

[4] Lindsay R, Burge RT, Strauss DM. One year outcomes and costs following a vertebral fracture. *Osteoporos. Int.* 2005;16:78–85.

[5] Jahelka B, Dorner T, Terkula R, et al. Health–related quality of life in patients with osteopenia or osteoporosis with and without fractures in a geriatric rehabilitation department. *Wien. Med. Wochenschr.* 2009;159:236-40.

[6] Rabenda V, Manette C, Lemmens R, et al. The direct and indirect costs of the chronic management of osteoporosis: a prospective follow- up of 3440 active subjects. *Osteoporos. Int.* 2006;17:1346–52.

[7] Yoshikawa T, Turner CH, Peacock M, et al. Geometric structure of the femoral neck measured using dual - energy x-ray absorptiometry. J Bone Miner Res 1994;9(7):1053-64. Erratum in: *J. Bone Miner. Res.* 1995;10(3):510.

[8] Nakamura T, Turner CH, Yoshikawa T, et al. Do variations in hip geometry explain differences in hip fracture risk between Japanese and white Americans? *J. Bone Miner. Res.* 1994;9:1071–6.

[9] Varga S: *Another View on the Fuzzy Regression.* Forum Statisticum Slovacum 2009;3:1-7.

[10] Kleinbaum DG, Klein M. Statistics for biology and health. 2nd ed, Springer *Science USA*, 2002. pp. 1-124.

[11] Varga Š. Fuzzy predictions in regression models *J. Appl. Mathem. Open. Access* 2010;3(1):245-251.

[12] Wendlová J. Regression model in estimate of femoral neck fracture by fall. *Open Access Emergency Medicine* 2010;2:29-36.

[13] Bergot C, Bousson V, Meunier A, Laval-Jeantet M, Laredo JD. Hip fracture risk and proximal femur geometry from DXA scans. *Osteoporos. Int.* 2002;13(7):542-50.

[14] Brownbill RA, Lindsey C, Crncevic-Orlic Z, et al. Dual hip bone mineral density in postmenopausal women: geometry and effect of physical activity. *Calcif. Tissue Int.* 2003;73(3):217-24.
[15] Im GI, Lim MJ. Proximal hip geometry and hip fracture risk assessment in a Korean population. *Osteoporos. Int.* 2011;22(3):803-7.
[16] Lektrakul N, Ratarasarn O. Geometry of proximal femur in the prediction of femoral neck fracture in the elderly female Thai population. *J. Med. Assoc. Thai.* 2009;92(Suppl5):S60-6.
[17] Ito M, Wakao N, Hida T, et al. Analysis of hip geometry by clinical CT for the assessment of hip fracture risk in elderly Japanese women. *Bone* 2010;46(2):453-7.
[18] Ulusov H, Bilgici A, Kuru O, el al. A new value of proximal femur geometry to evaluate hip fracture risk: true moment arm. *Hip. Int.* 2008;18(2):101-7.
[19] Gregory JS, Testil D, Stewart A, et al. A method for assessment of the shape of the proximal femur and its relationship to osteoporotic hip fracture. *Osteoporos. Int.* 2004;15:5–11.
[20] Alonso Gomez C, Curiel Diaz M, Carranza Hawkins F, et al. Femoral bone mineral density, neck shaft angle and mean femoral neck with as predictors of hip fracture in men and women. *Osteoporos. Int.* 2000;11:714–20.
[21] El Kaissi S, Pasco JA, Henry MJ, et al. Femoral neck geometry and hip fracture risk: the Geelong osteoporosis study. *Osteoporos. Int.* 2005;16:1299–303.
[22] Gnudi S, Malavolta N, Testi D, et al. Differences in proximal femur geometry distinguish vertebral from femoral neck fractures in osteoporotic women. *The Brit. J. Radiol.* 2004;77:219–23.
[23] Gnudi S, Ripamonti C, Lisi L, et al. Proximal femur geometry to detect and distinguish femoral neck fractures from trochanteric fractures in postmenopausal women. *Osteoporos. Int.* 2002;13:69–73.
[24] Faulkner KG, Wacker WK, Barden HS, et al. Femur strength index predicts hip fracture independent of bone density and hip axis length *Osteoporos. Int.* 2006;17:593–9.
[25] Wendlová J: Expected frequency of femoral neck fractures by fall in the osteoporotic and osteopenic East Slovak female population (Epidemiological Study). *Wien. Med. Wschr.* 2010;160(15-16):399-405.

[26] Wendlová J. Why is so important to balance the muscular dysbalance in mm. coxae area in osteoporotic patients? *Bratisl. lek. listy* 2008;109:502-7.

[27] Pahr DH, Zysset PK. From High-Resolution CT Data to Finite Element Models: Development of an Integrated Modular Framework; *Comput. Math. Biomech. Biomed. Engin.* 2009;12: 45-57.

Chapter IV

Femur Strength Index Versus Bone Mineral Density: New Findings *(Slovak Epidemiological Study)*

Abstract

Patients and Methods: We analysed the data in the sample (n = 3,215) of East Slovak women with a primary or secondary osteopenia, osteoporosis and with risk factors for osteoporosis, aged 20 –89 years, $\widetilde{x}$ = 59 years, 95% C. I. (59.31; 60.07). obtained from dual energy X-ray absorptiometry device (Prodigy – Primo, GE, USA). Measured variables: 1. left proximal femur: *T – score total hip, FSI (femur strength index),* 2. lumbar vertebrae L1–L4: *BMD (bone mineral density).*

Objectives: 1. To estimate and to compare an expected frequency of pathological *FSI < 1* and T – *score total hip ≤ - 2.5 SD* values in the East Slovak female population. 2. To estimate expected frequency of women with: *FSI < 1* and T – *score total hip ≤ - 2.5 SD* (Group A)*, FSI < 1* and *T – score total hip* from interval ⟨ *-1,0; - 2,5) SD* (Group B), *FSI < 1* and *T – score total hip > - 1, 0 SD* (Group C) in the East Slovak female population. 3. To determine, if *FSI* variable value is a significant predictor of *BMD* variable values in lumbar vertebrae.

Results: 1. In the East Slovak female population we can expect 14, 54% of women with *FSI* values < 1 and 6, 25% of women with osteoporosis in the total hip area according to *T – score.* 2. For the group A we can expect the mean value (μ) from interval (1,41; 2,36)%, for the group B from interval (4,50;6,03)% and for the group C from interval

(6,76;8,55)%. 3. Between *FSI* and *BMD* L1–L4 variable values there is not a statistically significant dependence, because *FSI* variable is quantitative and qualitative different variable from *BMD* variable.

Conclusion: The measurement of *FSI* variable values may discover a higher percentage of women with a probability of femoral neck fracture by fall than the measurement of *BMD* variable value in the total hip area. Patient with osteopenia or normal *BMD* measured in the total hip area may sustain a femoral neck fracture by fall, when she has pathological value of *FSI*, i.e. she has adverse values of geometric variables of proximal femur (*biomechanically unfavourable proximal femur configuration*). *FSI* variable value is not a significant predictor of *BMD* variable values in lumbar vertebrae L1–L4.

Keywords: osteoporosis - femur strength index (FSI) -- bone mineral density (BMD) – femoral neck fracture – dual energy X-ray absorptiometry (DXA)- lumbar vertebrae

INTRODUCTION

With the development of clinical osteology, spheres of interest, objectives and directions of epidemiological studies have been changing. Rising costs of complex treatment of fractures and social care of patients with permanent consequences of fractures (reduced work skills, reduced self-sufficiency, disability) brought to attention the precise assessment of bone quality, fracture risk and the prevention strategy of osteoporotic fractures [1,2] . The treatment costs of fractures caused by falls represent still an economic burden for the health care system [3–7]. Therefore, the search continues to find more precise variables in densitometric methods (DXA and QCT – quantitative computed tomography) to determine patients with a fracture risk by fall.

In this clinical study we were interested in comparison of two densitomeric variables measured by DXA and their possibility to discover the female patients at high risk of proximal femoral fracture by fall.

OBJECTIVES OF THE STUDY

1. To estimate an expected frequency of the occurrence of pathological *FSI* <1 variable values and pathological *BMD (T –score total hip* ≤ -

2.5 SD) variable values in the East Slovak female population and to compare them.

2. In the East Slovak female *population* to estimate expected 95% C.I. for percentage of women with variable values:

- FSI < 1 and T – score total hip ≤ - 2.5 SD, Group A
- FSI < 1 and T – score total hip from interval 〈 -1.0; - 2.5) SD, Group B
- FSI < 1 and T – score total hip > - 1. 0 SD, Group C: and in the sample to estimate the percentage of women belonging into groups A, B or C.

3. To determine, if FSI variable value is a significant predictor of BMD values in lumbar vertebrae L1 –L4

PATIENTS AND METHODS

Characteristics of the Sample

Using a DXA bone densitometer (Prodigy – Primo, GE, USA) we analysed the data in the sample of East Slovak women (*n* = 3,215) aged 20 – 89 years, $\widetilde{x}$ = 59 years, 95% C. I. (59.31; 60.07):

1. without case – history of femoral neck fracture
2. with risk factors for the development of osteoporosis
3. with a primary or secondary osteopenia
4. with a primary or secondary osteoporosis

All women were examined with the same bone densitometer DXA. The *BMD* was determined in the standard region of interest (ROI) – total hip left. *BMD* values were given in absolute numbers in *g of Ca-hydroxyapathite crystals for* cm^2 *(*g/cm^2*)*, as well as in relative numbers as T-score (the number of standard deviations from the reference group of young healthy women). Osteoporosis or osteopenia were diagnosed in accordance with the WHO criteria.

To achieve the measurement quality (QA), only two operators alternated in measuring with the DXA device and all women were measured with the same device. The following variables were measured:

1. Proximal femur left: BMD total hip (T-score total hip), FSI
2. Lumbar vertebrae L1 –L4: BMD

Characteristics of the population: all women of the East Slovakia with osteopenia, osteoporosis and with risk factors for osteoporosis (hereafter East Slovak female population).

Definitions of Measured Variables and Variables Describing in the Discussion [8]

FSI (femur strength index) – is a biomechanical variable determining whether the bone strength in the femoral neck area endures the load of compressive force impact by fall (normal value: $FSI \geq 1$, pathological value: $FSI < 1$).

It is defined as a ratio of estimated elastic limit in compression of the femoral neck (σ_E) to the expected compressive stress of a fall on the greater trochanter adjusted for the patient´s age, height and weight (σ_C) .

Alpha angle (α) is an angle formed by the femoral shaft axis and the perpendicular. The alpha angle can acquire both positive and negative values in the population, depending whether the femur is in a valgus or varus position.

Theta angle (θ) is an angle formed by the femoral neck axis and the femoral shaft axis.

HAL (hip axis length) -- is a distance (in mm) from the beginning point of the greater trochanter protuberance to the pelvis inner rim, measured in the femoral neck axis.

CSMI (cross sectional moment of inertia) – is defined as the sum of multiplications of elementary areas and the squares of their distances from neutral axis, denoted as I_Y (given in cm^4). The larger the cross sectional area, the higher the number of small elementary areas (A) and so is the bigger the second root of the distance (z^2) of these elementary areas from the neutral axis. The enlargement of the cross sectional area is accompanied by the enlargement of the cross sectional moment of inertia.

STATISTICAL ANALYSIS [9-13]

To analyse the data of the sample, statistical methods were applied using statistical programme systems Statgraphics Centurion XV.

1. We tested the character of the distribution of *FSI*, *T –score total hip* and *BMD L1 –L4 variable* values in the sample.
2. Using the Goodness-of-Fit Test χ^2 (chi-Squared Test) and coming from the character of empirical distribution of the frequencies (probability) of *FSI* variable values in the *sample,* we estimated the expected frequencies of *FSI* variable values in *the East Slovak female population* at the significance level $\boldsymbol{\alpha = 0.01}$ We tested the character of the distribution of variable values in the sample and using the Goodness-of-Fit Test χ^2 (chi-Squared Test) and coming from the character of empirical distribution of the frequencies (probability) of *T-score total hip* variable values in the *sample,* we estimated the expected frequencies of *T-score total hip* variable values in the *East Slovak female population* at the significance level $\alpha = 0.05$.
3. In the sample we calculated percentage of women, which belong into group A, B or C.
4. In the East Slovak female population, we calculated the 95% C.I. (confidence interval) for expected percentage frequencies of women, which belong into group A, B or C.
5. Linear regression analysis: To estimate the Pearson′s correlation coefficients between *FSI* and *BMD* L1–L4 variable values. The coefficients verify the linear association and measure the intensity of association between variables *FSI* and *BMD* L1 –L4.
6. To illustrate the results we used the following statistical graphs:

   - histograms, circle diagram

RESULTS

Age Variable

The median $(\widetilde{x})$ for age is 59 years, 95% C. I. (59.31; 60.07).

FSI Variable

The Table 1 brings an empirical distribution of *FSI* variable values frequencies in the sample and the expected distribution of the *FSI* values frequencies in the East Slovak female population.

Table 1. Observed and expected frequencies of *FSI* variable values in East Slovak female population. (Chi-Squared Test, Goodness-of-Fit Test)

	FSI					
	Lower Limit	*Upper Limit*	*Observed Frequency*	*Expected Frequency*	*Expected Freq. in %*	*Chi-Squared*
at or below		0.5	6	5.77	0.18	0,01
	0.5	1.0	465	461.44	14.36	0,03
	1.0	1.5	1614	1644.68	51.17	0.57
	1.5	2.0	859	812.14	25.27	2.70
	2.0	2.5	191	202.48	6.30	0.65
	2.5	3.0	45	55.46	1.73	1.97
	3.0	3.5	17	18.37	0.57	0.10
	3.5	4.0	6	7.12	0.22	0.18
	4.0	4.5	3	3.12	0.10	0.00
above	4.5		8	3.42	0.11	6.13

Chi-Squared = 12. 3499 with 6 d.f. P-Value = 0.0546021.
Goodness-of-Fit Tests, Chi-Squared Test.
Sample: n = 3, 214 (from the sample n = 3, 215 one error value FSI = 0 was excluded).

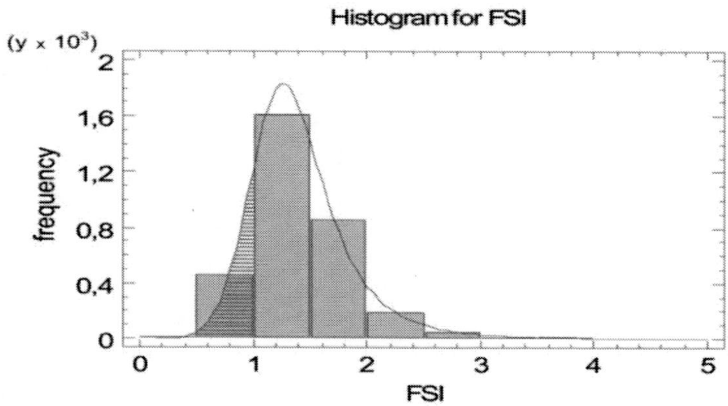

Figure 1. Expected distribution of frequencies for *FSI* variable values in the East Slovak population of osteopenic , osteoporotic women and women with risk factors for osteoporosis. Hatched field represents an expected number of frequencies for *FSI < 1.*

In the East Slovak female population we can expect 0.18% of women with the *FSI* values < 0.5 and 14.36% of women with the *FSI* values from the interval (0.5;1.0). It means that we can expect 14.36% + 0.18% = 14.54% of women in the East Slovak female population with *FSI* values < 1. At the significance level $\alpha = 0{,}01$ it can be assumed that the probability distribution for *FSI* variable values in the East Slovak female population is loglogistic with the parameters derived on the basis of values found in the sample. One patient with faulty measurement *FSI =0* was excluded from sample n = 3214 (Figure 1).

T-Score Total Hip Variable

In the East Slovak female population according to values measured in the total hip area we can expect 6.25% of osteoporotic women with the T-score values ≤ 2.5 SD and 34.94% of osteopenic women with T-score values from the interval ⟨ -1.0; - 2.5) *SD.* At the significance level $\alpha = 0{,}05$ it can be assumed that the probability distribution for *T-score total hip* variable values in the East Slovak female population is loglogistic with the parameters derived on the basis of values found in the sample (Table 2, Figure 2).

The expected frequency of the incidence of pathological *FSI* values of the East Slovak female population is 2.33 times higher as the expected frequency of the incidence of pathological *T - score* for osteoporosis measured in the total hip area (Table 3).

Table 2. Observed and expected frequencies of *T – score total hip* variable values in East Slovak female population

T-score total hip left (SD)	Observed frequency in the sample (total numbers)	Observed frequency in the sample in %	Expected frequency in population in %
(- ∞; - 2.5⟩	196	6.10	6.25
(- 2.5; - 1.0⟩	1142	35.52	34.94
(- 1.0; 0.0⟩	1089	33.87	34.59
(0.0;+ ∞)	788	24.51	24.22

Sample: n= 3,215.

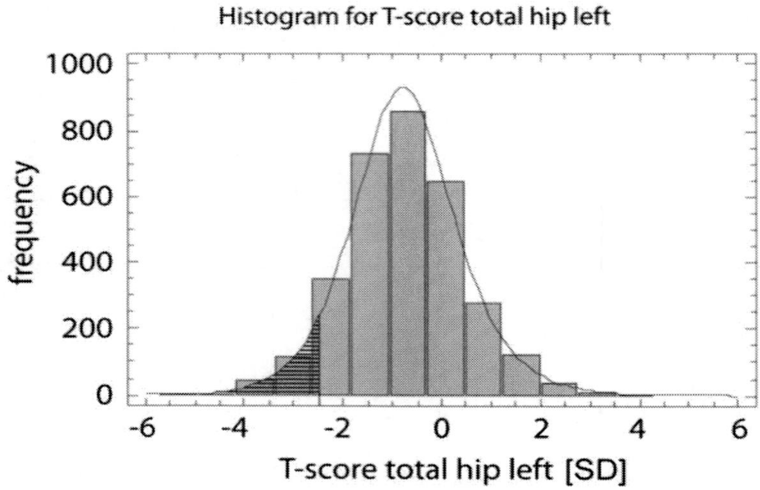

Figure 2. Expected distribution of frequencies for *T – score total hip left* variable values in the East Slovak population of osteopenic , osteoporotic women and women with risk factors. Hatched field represents an expected number of frequencies for *T – score ≤ - 2.5 SD.*

Table 3. Comparison of expected frequency of pathological *FSI* variable values and *T – score total hip* values in East Slovak female population

Variable	Observed frequency in the sample (total numbers)	Expected frequency in population (%)
FSI left < 1	471	14.54
T–score ≤ - 2.5 SD (total hip left)	196	6.25

FSI and T - Score Total Hip Variables

In the sample, from total number of women with FSI < *1* (n = 471), there were 51.91% women with *FSI < 1* and with *T – score total hip > - 1. 0 SD,* 35.59% women with *FSI < 1* and with *T – score total hip* from interval ⟨ *-1.0; - 2.5) SD* and 12.50% women with *FSI < 1* and with *T – score total hip ≤ - 2.5 SD* (Figure 3, Table 4b).

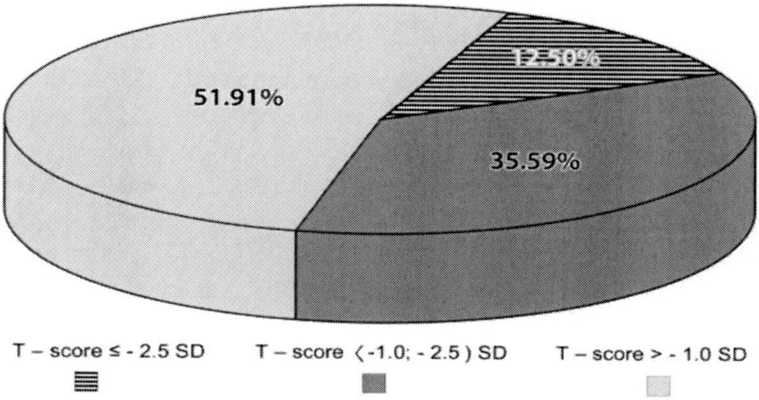

Figure 3. Observed frequencies of *T – score total hip left* variable values in percentage for female patients with *FSI left < 1* in the sample (n = 471).

Table 4a. Frequencies of *FSI* variable values in the sample

Sample: n = 3214		
Variable FSI	Total frequency	Frequency in %
FSI<1	471	14.65
FSI ≥1	2743	85.34

Table 4b. Observed and expected frequencies of categorized *FSI* and *T –score total hip*

Observed Frequencies in the Sample and Expected Frequencies in the Population					
Variable		T-score ≤ -2.5 SD	T-score ⟨ -1. 0; -2.5) SD	T-score > -1.0 SD	Total
	Observed frequencies	59	168	244	471
FSI<1	Observed frequencies in %	1.84	5.23	7.62	14.65
	Expected frequencies in % (95% C.I.)	1.41-2.36	4.50-6.03	6.76-8.55	-----

In the sample from total number of women (n = 3215) there were 7.62% women with *FSI* < *1* and with *T – score total hip* > *- 1. 0 SD, 5.23%* women with *FSI* < *1* and with *T – score total hip* from interval 〈 *-1.0; - 2.5) SD* and 1.84% women with *FSI* < *1* and with *T – score total hip* ≤ *- 2.5 SD* (Table 4b).

In the East Slovak female population it can be expected with the probability of 0.95 (95%), that the mean value (μ) of percentage (%) will be from:

- interval (1.41;2.36) for women with *FSI* < *1* and simultaneously with *T – score total hip* ≤ *- 2.5 SD*
- interval (4.5;6.03) for women with *FSI* < *1* and simultaneously with *T – score total hip* from interval 〈 *-1.0; - 2.5) SD*
- interval (6.76;8.55) for women with *FSI* < *1* and simultaneously with *T– score total hip* > *- 1, 0 SD* (Tab 4a, 4b).

FSI AND BMD L1– L4 VARIABLES

Linear regression analysis: At the significance level α = 0. 05 there is a very low direct dependence between *FSI* and *BMD* L1 –L4, the Pearson´s correlation coefficients are near to the zero. Variable FSI explains the variability of BMD L1 variable values only in 0.250%, of BMD L2 only in 0.230%, of BMD L3 only in 0.380% and of BMD L4 in less than 0.0% (Table 5).

Table 5. Pearson´s coorelation coefficinets (k) for linear direct dependence between *FSI* and *BMD* L1 –L4 variable values at the significance level α = 0,05 (n = 3, 353)

Lumbar vertebrae	Pearson´s correlation coefficients (k)	p - value
L1	0.05000	0.0038*
L2	0.04858	0.0049*
L3	0.06168	0.0004*
L4	0.03348	0.0526

Discussion

At present, *FSI* is one of a few available variables, corresponding to biomechanical criteria of loading the bone by fall, whose values can be determined by DXA in a routine ambulatory practice. *FSI* integrates in itself three important bone characteristics:

1. BMD
2. bone geometry: HAL, angle alpha (α), angle theta (θ), CSMI in the minimum cross sectional area of femoral neck (see definitions p. 4)
3. elasticity and strength

Whether the *FSI* variable value is calculated from the two-dimensional bone measurement by DXA, which in comparison with three-dimensional measurement has some inaccuracies, it is a value determining the bone quality much more precisely than *BMD*. Clinical importance of the *FSI* variable value lies in the fact that it enables to determine an individual forecast for femoral neck fractures by fall. The forecast of femoral neck fractures by fall on the basis of *BMD* is markedly limited, as *BMD* characterises only one physical value – density, which determine only a part of bone quality.

Authors of some clinical retrospective or prospective studies [14 – 21] agree that the increase of the values of geometric variables of proximal femur • *HAL*, • *theta angle* , • *d* (average width of femoral neck) over the mean value in the population is a risk factor for femoral neck fractures.

Faulkner et al. [22] compared *BMD, HAL, and CSMI* in femoral neck area and *FSI* obtained from DXA measurements in a group of women with and without hip fracture. Femoral neck *BMD* and *FSI* were significantly lower and *HAL* significantly higher in fractured group compared with controls. Mean *CSMI* was not significantly different between fracture patients and controls. After adjustment of these variables in regression model, the authors conclude that *BMD, HAL* and *FSI* are significant independent predictors of hip fracture. Crabtree et al [23] concluded in their work, that the geometric variation of proximal femur may contribute to the large variations in hip fracture risk across Europe.

In our clinical study we were interested in percentage comparison of pathological FSI < 1and T – score total hip ≤ - 2,5 SD variable values in the same sample of East Slovak women and to estimate the expected frequencies of this variable values in the East Slovak female population. The expected percentage of *FSI* pathological variable values occurrence (14.54%) is 2.33 times higher than the expected percentage of osteoporosis occurrence in the total hip area (T-score ≤ - 2.5 SD) in the East Slovak female population.

It is evident from such a simple comparison that *BMD* and *FSI* variables are two different characteristics of bone quality. *FSI,* which characterizes the bone quality by means of several integrated biomechanical variables, including in itself the *BMD* variable, discovers a higher percentage of women at risk for femoral neck fracture by fall than *BMD.* The results show that even a patient with osteopenia or normal *BMD* measured in the total hip area may sustain a femoral neck fracture by fall, in case she has pathological value of *FSI* variable, i.e. she has adverse values of geometric variables of proximal femur *HAL, angle α, angle θ* (so – called *biomechanically unfavourable proximal femur configuration*).

In our previous published studies [24-25], we have estimated from predictive logistic regression model, created from East Slovak female data, that the rise of *HAL, angle α, angle θ* variable values by one unit statistically significant raises the odds for femoral neck fracture by fall. It verifies that the variation of *HAL, angle α, angle θ* variable values significant influences the risk of femoral neck fracture by fall.

The measurement of *FSI* value in clinical practice enables to improve the strategy of treatment in the prevention of femoral neck fractures.

The *FSI* value < 1 could therefore be used as a criterion for the initiation of preventive therapeutic interventions in order to avoid femoral neck fractures. These interventions may include:

1. drug therapy
2. wearing a hip joint protector
3. kinesitherapy:
    - coordination exercises
    - exercises to remove the muscular dysbalance in the mm. coxae area [25]
    - training of techniques how to fall correctly

FSI variable is quantitative and qualitative different variable from *BMD* variable, what shows a little portion of variability of *FSI* variable values on the variability of *BMD* L1 –L4 variable values according to linear regression analysis in our study. Prediction of *BMD* variable values by the help of *FSI* variable values is unreliable and therefore impossible in practice.

Methods of measurement and determination of bone quality have been constantly improving. As demonstrated by the latest clinical studies, the future of bone densitometry lies in a three-dimensional measurement of *FSI* an other new progressive biomechanical variables and also in high resolution QCT [27 – 31].

Conclusion

New types of DXA and QCT densitometers provide programs for clinical and ambulatory practice enabling to measure different biomechanical and geometric variables of proximal femur. Regrettably, these programs are utilized minimally in practice and osteologists are still rigidly adhering to *BMD* values. Offered programs overrun the "guidelines" of international and national osteological societies, which still have not included the criteria for the evaluation of new biomechanical variables.

Summing Up

1. The measurement of *FSI* variable values may discover a higher percentage of women in the population with a risk of femoral neck fracture by fall than the simple measurement of *BMD* variable values in the total hip area.
2. Patient with osteopenia or normal *BMD* measured in the total hip area may sustain a femoral neck fracture by fall, when she has pathological *FSI* value i.e. she has adverse values of geometric variables of proximal femur (so – called *biomechanically unfavourable proximal femur configuration*).
3. Between *FSI* and *BMD* L1 –L4 variable values there is not a statistically significant direct dependence, because *FSI* variable is quantitative and qualitative different variable from *BMD* variable. *FSI* variable value cannot be a predictor of the *BMD* variable values in lumbar vertebrae L1-L4.

4. The study investigated East Slovak women. Although it can be supposed that similar results will be found in the whole Slovak female population and in other female populations at least in central Europe, further studies will have to verify this.

Declaration – Conflict of Interests

The author declares, that she has no competing interests (financial or non financial).

References

[1] Pietschmann P, Kerschan – Schindl K : Knochenqualität – wissenschaftliche Aspekte versus praktische Relevanz. *J. Miner. Stoffwechs*. 2004;11:16–18.

[2] Boonen S, Singer AJ: Osteoporosis management: impact of fracture type on cost and quality of life in patient at risk of fracture. *Curr. Med. Res. Opin*. 2008;24:1783–1790.

[3] Stein VK, Dorner TH, Lawrence K et al: Economic concepts for measuring the costs of illness of osteoporosis: An international comparison. *Wien. Med. Wochenschr*. 2009;159:253-261.

[4] Finnern HW, Sykes DP: The hospital cost of vertebral fractures in the EU: estimates using national data sets. *Osteoporos. Int*. 2003;14:429–436.

[5] Lindsay R, Burge RT, Strauss DM: One year outcomes and costs following a vertebral fracture. *Osteoporos. Int*. 2005;16: 78 – 85.

[6] Jahelka B, Dorner T, Terkula R et al: Health–related quality of life in patients with osteopenia or osteoporosis with and without fractures in a geriatric rehabilitation department. *Wien. Med. Wochenschr*. 2009;159:236-240.

[7] Rabenda V, Manette C, Lemmens R et al.: The direct and indirect costs of the chronic management of osteoporosis: a prospective follow-up of 3440 active subjects. *Osteoporos. Int*. 2006;17:1346–1352.

[8] Nakamura T, Turner CH, Yoshikawa T et al: Do variations in hip geometry explain differences in hip fracture risk between Japanese and white Americans? *J. Bone Miner. Res*. 1994;9:1071–1076.

[9] Rublíková E, Labudová V, Sandtnerová S: *Analysis of categorical data.* (Analýza kategoriálnych údajov), 1st Edition, Publishing house Economic University, Bratislava, 2009, pp. 41-141.

[10] Pacáková V: *Statistical methods for economists*, 2nd Edition, Publishing House: Iura Edition, Bratislava, 2009, pp. 174–7.

[11] Varga S: *Another View on the Fuzzy Regression.* Forum Statisticum Slovacum 2009; 3:1–7.

[12] Pacáková, V.: *Aplikovaná poistná štatistika* (Applied Actuarial Statistics), 1st Edition, Publishing House: IURA Edition Bratislava , 2004,pp. 67-86.

[13] Varga Š: Fuzzy predictions in regression models. *J. Appl. Mathem. Open Access* 2010;3 (1):245-251.

[14] Kukla C, Gaebler C, Pichl RW et al.: Predictive geometric factors in a standardized model of femoral neck fracture. Experimental study of cadaveric human femurs. Injury, *Int. Care Injured* 2002;33:427–433.

[15] Gregory JS, Testil D, Stewart A et al.: A method for assessment of the shape of the proximal femur and its relationship to osteoporotic hip fracture. *Osteoporos. Int.* 2004;15:5–11.

[16] Alonso Gomez C, Curiel Diaz M, Carranza Hawkins F et al.: Femoral bone mineral density, neck shaft angle and mean femoral neck with as predictors of hip fracture in men and women. *Osteoporos. Int.* 2000;11:714–720.

[17] El Kaissi S, Pasco JA, Henry MJ et al.: Femoral neck geometry and hip fracture risk: the Geelong osteoporosis study. *Osteoporos. Int.* 2005;16:1299–1303.

[18] Watts NB: Fundamentals and pitfalls of bone densitometry using dual – energy X – ray absorbtiometry (DXA). *Osteoporos. Int.* 2004;15:847–854.

[19] Gnudi S, Malavolta N, Testi D, Viceconti M: Differences in proximal femur geometry distinguish vertebral from femoral neck fractures in osteoporotic women. *The Brit. J. Radiol.* 2004;77: 219 – 223.

[20] Wu SY, Qi J, Lu Y et al: Densitometric and geometric measurement of the proximal femur in elderly women with and without osteoporotic vertebral fractures by volumetric quantitative multi-slice CT. *J. Bone Miner. Metab.* 2010;25(4):521-9..

[21] Gnudi S, Ripamonti C, Lisi L et al.: Proximal femur geometry to detect and distinguish femoral neck fractures from trochanteric fractures in postmenopausal women. *Osteoporos. Int.* 2002;13:69–73.

[22] Faulkner KG, Wacker WK, Barden HS et al.: Femur strength index predicts hip fracture independent of bone density and hip axis length. *Osteoporos. Int.* 2006;17:593–599.

[23] Crabtree N, Lunt M, Holt G et al: Hip geometry, bone mineral distribution, and bone strength in European men and women: The EPOS study. *Bone* 2000;27:151-9.

[24] Wendlova J: Logistic regression in estimate of femoral neck fracture by fall *Open Access Emergency Medicine* 2010;2:29-36.

[25] Wendlova J: Expected frequency of femoral neck fractures by fall in the osteoporotic and osteopenic East Slovak female population. (Epidemiological Study).*Wien. Med. Wschr.* 2010;159: in press.

[26] Wendlová J: Why is so important to balance the muscular dysbalance in mm. coxae area in osteoporotic patients? *Bratisl. lek. listy* 2008;109:502–507.

[27] Cheng X, Li J, Lu Y, Keyak J, Lang T: Proximal femoral density and geometry measurements by quantitative computed tomography. Association with hip fracture. *Bone* 2007;40:169–174.

[28] Manske SL, Liu Ambrose T, De Bakker PM et al: Femoral neck cortical geometry measured with magnetic resonance imaging is associated with proximal femur strength. *Osteoporos. Int.* 2006;17:1539–1545.

[29] Bousson V, Le Brass A, Roqueplan F et al: Volumetric quantitative computed tomography of the proximal femur: relationships linking geometric and densitometric variables to bone strength. Role for compact bone. *Osteoporos. Int.* 2006;17:855–864.

[30] Engelke K, Adams JE, Armbrecht G et al.: Clinical Use of Quantitative Computed Tomography and Peripheral Quantitative Computed Tomography in the Management of Osteoporosis in adults. The 2007 ICSD Official Position. *J. Clin. Densit.* 2008;11:123-162.

[31] Sipos W, Pietschmann P, Rauner M et al: Pathophysiology of osteoporosis. *Wien. Med. Wochenschr.* 2009;159:230-234.

Chapter V

T – Plus Z – Score in Assessment of Fracture Risk in Female Population from Bratislava *(Epidemiological Study)*

Abstract

Patients and methods: We analysed 498 (n = 498) densitometric findings from the area of proximal femur: Neck (ROI_1), Ward ´s area (ROI_2), great trochanter (ROI_3) (DXA - dual energy X–ray absorptiometry, Osteocore II, France) from the population of Bratislava women of the average age of 59.71 years ⟨58.63; 60.80⟩. Measured values of T and Z– score were categorized as follows: 1 – increased fracture risk (T – score < - 2.5 SD/ T = 1, Z–score ≤ - 1.0 SD/ Z = 1), 0 – without increased fracture risk (T – score > - 2.5 SD/ T = 0, Z – score > - 1.0 SD / Z = 0).

The objective of the study: We aimed to prove that the relative risk of fractures (RR) is higher in women whose T and Z–score values come under the category of increased risk than in the women whose T–score only falls within the category of increased risk.

Statistical analysis: 1. Contingency tables, measurement of the intensity of the association of dichotomous variables 2. Determination of probability (P) and RR.

Results: Statistically significant intensity of the association of dichotomous variables was proved for measurements in ROI_1, ROI_2, and

ROI_3. The closest association (0.797) for dichotomous variables was found in ROI_2 (Ward's area). RR for development of fracture was higher for all measurement sites (ROI_1 – 29.87; ROI_2 – 15.3; ROI_3 – 5.5). When the women were divided according to age into two groups: a) women< 65 years, b) women ≥ 65 years, it was found that the RR of fracture for women < 65 years having T – score < - 2.5 SD and also Z – score ≤ - 1.0 SD is very high (RR = ∞), old women (≥ 65 years) had the RR for ROI_1 – 32.61, for ROI_2 – 7.72, and for ROI_3 – 6.9. Probability (P) of the presence of women with T – score values < - 2.5 SD and Z – score ≤ - 1.0 SD in the group under 65 years is for ROI_1 – 12.5%; ROI_2 – 34.4%; ROI_3 – 20.5%, and the probability for women ≥ 65 years is for ROI_1 – 75%; ROI_2 – 98%; and ROI_3 – 100%.

Conclusions: The highest intensity of the association of dichotomous variables in Ward's area is conditional on measuring greater fraction of the trabecular bone in this area in comparison with ROI_1 and ROI_3, and so measurement errors resulting from osteoarthrotic changes in cortical bone in the measured area are less significant. These changes are reflected in the dispersion of T and Z – score values and in their weaker association for ROI_1 and ROI_3. The study's contribution lies in the finding that the relative fracture risk in proximal femur in women increases, if T-score < - 2.5 SD and, simultaneously, Z-score ≤ - 1.0 SD, and this risk is especially high for women under the age of 65. In assessing the individual fracture risk (by questionnaires) in female patients we recommend to add also Z – score values in the measured area to T –score values and to other anamnestic and biochemical risk factors.

Keywords: total hip - osteoporosis – female population – T- score – Z – score – fracture risk

Introduction

The assessment of individual fracture risk has been constantly improved and made more accurate. The assessment of individual fracture risk by applying only the T – score values does not take into consideration other risk factors, and is, therefore, distorted. For example, in the past the same fracture risk was assessed for T – score = - 2.6 SD in a woman of 45 as well as a woman of 65, without considering the age factor [1, 2, 3]. This was the reason why other risk factors began to be added to T – score values, as age, sustained fracture after the age of 50, femur neck fracture in mother, increased biochemical markers of osteoresorbtion, smoking, body mass under 56 kg, and the like. These risk factors in the individual assessment of fracture risk are evaluated together with T – score from the area of proximal femur (total hip) using various questionnaires with good sensitivity and specificity, contributing to the objectification of the risk assessment [4].

Z – score determination is a part of all densitometric measurements and it is given in all densitometric findings. Its significance in the assessment of individual fracture risk is defined as a risk increase with the decrease of standard deviation (SD) by every -1.0 SD. However, in practice it is not used for individual fracture assessment at all. An important finding is that in postmenopausal women with reduced Z-score values it is necessary to consider the presence of secondary osteoporosis.

We have set up the hypotheses in the study in order to objectively prove the significance of Z– score values in the individual fracture risk assessment and to introduce it into the medical practice

The Objective of the Study

1. To establish how many times the relative fracture risk is higher in the group of women with T-score < - 2.5 SD and, simultaneously, with Z-score ≤ - 1.0 SD, in comparison with women with T-score < - 2.5 SD and Z-score > - 1.0 SD.
2. To divide the cohort of women into two groups according to age:

- age ≥ 65 years (old age)
- age < 65 years

To establish how many times the relative fracture risk is higher for women ≥ 65 years with T-score < - 2.5 SD and, simultaneously, with Z-score ≤ - 1.0 SD, in comparison with women ≥ 65 years with T-score < - 2.5 SD and Z-score > - 1.0 SD.

To establish how many times the relative fracture risk is higher for women < 65 years with T-score values< - 2.5 SD and, simultaneously, with Z-score ≤ - 1.0 SD, in comparison with women < 65 years with T-score < - 2.5 SD and Z – score > - 1.0 SD.

Characteristics of the Sample of Examined Female Patients and Methods Applied

We analysed 498 (n = 498) densitometric findings from the area of proximal femur in the sample of population of Bratislava women of the average age of 59.71 ⟨58. 63, 60. 80⟩. .

In the group of women from different Bratislava districts there were patients with osteoporosis and osteopenia who underwent control BMD measurement, as well as women with osteoporosis risk factors who did not undergo densitometric examination. The women with osteoporosis risk factors were selected in accordance with the indication list of the International Society for Clinical Densitometry (Official Positions 2004) (5). Within the group of risk women, new patients were diagnosed with osteoporosis and osteopenia. All women were examined with the same densitometer DXA (dual energy X–ray absorptiometry) Osteocore II, France. Bone mineral density (BMD) was determined in standard regions of interest for Osteocore II. (ROI - region of interest) – Neck (ROI_1), Ward's area (ROI_2), great trochanter (ROI_3). BMD values were given in absolute numbers in g of Ca- hydroxyapathite crystals for cm^2 (g/cm^2), as well as in relative numbers as T-score (the number of standard deviations from the reference group of young healthy women) and Z-score (the number of standard deviations from the relevant age group of healthy women). Osteoporosis was diagnosed in accordance with the WHO criteria:

Statistical Analysis [6-8]

We have used the statistical program systems Statgraphics plus version 3 and SPSS version 6 for the analysis of the female group data.

1) Test for normality of distribution of T – score and Z - score values in ROI_1, ROI_2, ROI_3 (χ^2 – goodness-of-fit statistic, Kolmogorov – Smirnov statistics)
2) Contingency tables (association)
   We have the T and Z – scores categorized as:

a)Presence of higher fracture risk (designation of category - 1)

b)for T – score < - 2.5 SD and for Z – score ≤ - 1.0 SD
c)(categorized T = 1, Z = 1)
d)Absence of higher fracture risk (designation of category – 0)
e)for T – score > - 2.5 SD and for Z – score > - 1.0 SD.
f)(categorized T = 0, Z = 0)

3) Verification of association and measurement of the intensity of association between categorized variables T and Z – score in contingency tables:

a)Phi coefficient
b)Cramer's V coefficient
c)Pearson's contingency coefficient

4) Assessment of relative fracture risk in the sample (6) according to T - score with regard to Z – score values

Because the incidence of osteoporosis in Slovakia is according to EVOS study less than 10% [EVOS, 9], in this retrospective study we can use RR instead the OR [10].

Results

Measured values of T – score in ROI_1 , ROI_2 have normal distribution, lognormal distribution of frequency is in ROI_3. Z– score values in ROI_1, ROI_2, ROI_3 are normally distributed.

Probability (P) and Relative Risk (RR) for ROI

ROI_1 (Table 1a,b)

$$P_1 = \frac{a}{a+b} = \frac{15}{20} = 0.25 = 25\%$$

$$P_2 = \frac{c}{c+d} = \frac{4}{478} = 0.00836 = 0.8\%$$

$$RR = \frac{0,25}{0.00836} = 29.875$$

ROI_2 (Table 2a, b)

$$P_1 = \frac{a}{a+b} = \frac{70}{111} = 0.631 = 63.1\%$$

$$P_2 = \frac{c}{c+d} = \frac{16}{378} = 0.0413 = 4.13\%$$

$$RR = \frac{0.631}{0.0413} = 15.3$$

ROI_3 (Table 3a, b)

$$P_1 = \frac{a}{a+b} = \frac{12}{43} = 0.279 = 27.9\%$$

$$P_2 = \frac{c}{c+d} = \frac{25}{455} = 0.055 = 5.5\%$$

$$RR = \frac{0.279}{0.055} = 5.07$$

Probability (P_1) that women in selected group have T – score < - 2.5 SD, if they have simultaneously Z – score ≤ - 1.0 SD, is for:

1) ROI_1 - 25%
2) ROI_2 – 63.1%
3) ROI_3 – 27.9%

Probability (P_2) that women in selected group have T – score < - 2.5 SD, if they have simultaneously Z – score > - 1.0 SD, is for:

1) ROI_1 - 0.8%
2) ROI_2 – 4.13%
3) ROI_3 – 5.5%

We defined RR for sustaining fracture according to T – score with regard to Z – score in ROI_1, ROI_2, ROI_3 as follows:

$$RR = \frac{P(T = 1 / Z = 1)}{P(T = 1 / Z = 0)}$$

Relative risk (RR) for sustaining fracture in the sample is for:

1) ROI_1 - 29.875 times
2) ROI_2 – 15.300 times
3) ROI_3 – 5.070 times

higher for women having T – score < - 2.5 SD and simultaneously Z – score ≤ - 1.0 SD than for the women having T – score < - 2.5 SD and simultaneously Z – score > - 1.0 SD.

In the population the RR is with probability 95% from the interval:

1) ⟨13.86; 64.39⟩ for ROI_1
2) ⟨11.22; 20.75⟩ for ROI_2
3) ⟨2.85; 9.04⟩ for ROI_3

There was a statistical significance of association intensity (Table 1b, 2b, 3b) for dichotomous variables in contingency tables (Table 1a, 2a, 3a) for all ROIs. The closest association was found for the contingency table Table 2a (ROI_2), r = 0.648 (Table 2b).

Table 1a. Contingency table with categorized T and Z - scores for ROI_1

	T – score (SD) ROI_1		Total count in row
Z – score (SD) ROI_1	> - 2.5 (count)	≤ - 2..5 (count)	
> - 1.0	474 d	4 c	478
≤ - 1.0	15 b	5 a	20
Total count in column	489	9	498

a, b, c, d, - frequencies.

Table 1b. Intensity of association for dichotomous variables in Table 1a (ROI_1)

Coeficient of assotiation	Intensity of assotiation	Approximate significance
Phi	0.35613	*
Cramer´s V	0.35613	*
Pearson´s	0.33549	*

$\alpha = 0.05$ $p > 0.05$ * statistical significance.

Table 2a. Contingency table with categorized T and Z - scores for ROI_2

	T – score (SD) ROI_2		Total count in row
Z – score (SD) ROI_2 (count)	> - 2.5 (count)	≤ - 2..5 (count)	
> - 1.0	371 d	16 c	387
≤ - 1.0	41 b	70 a	111
Total count in column	412	86	498

Table 2b. Intensity of association for dichotomous variables in Table 1a (ROI_2)

Coeficient of assotiation	Intensity of assotiation	Approximate significance
Phi	0.64885	*
Cramer´s V	0.64885	*
Pearson´s	0.54431	*

$\alpha = 0.05$ $p > 0.05$.

Table 3a. Contingency table with categorized T and Z - scores for ROI_3

	T – score (SD) ROI_3		Total count in row
Z – score (SD) ROI_3 (count)	> - 2.5 (count)	≤ - 2.5 (count)	
> - 1.0	430 d	25 c	455
≤ - 1.0	31 b	12 a	43
Total count in column	461	37	498

Table 3b. Intensity of association for dichotomous variables in Table 1a (ROI_3)

Coeficient of assotiation	Intensity of assotiation	Approximate significance
Phi	0.24004	*
Cramer´s V	0.24004	*
Pearson´s	0.23341	*

$\alpha = 0.05$ $p > 0.05$.

Probability (P) and Relative Risk (RR) for Women< 65 Years

ROI₁

$$P_1 = \frac{a}{a+b} = \frac{2}{16} = 0.125 = 12.5\%$$
$$P_2 = \frac{c}{c+d} = \frac{0}{306} = 0.0 = 0.0\%$$
$$RR = \frac{0.125}{0.0} = \infty$$

ROI₂

$$P_1 = \frac{a}{a+b} = \frac{21}{61} = 0.344 = 34.4\%$$
$$P_2 = \frac{c}{c+d} = \frac{0}{261} = 0.0 = 0.0\%$$
$$RR = \frac{0.344}{0.0} = \infty$$

ROI₃

$$P_1 = \frac{a}{a+b} = \frac{8}{39} = 0.205 = 20.5\%$$
$$P_2 = \frac{c}{c+d} = \frac{0}{283} = 0.0 = 0.0\%$$
$$RR = \frac{0.205}{0.0} = \infty$$

Probability (P_1) that women in selected group aged < 65 years have T – score < - 2.5 SD, if they have simultaneously Z – score ≤ - 1.0 SD is for:

1) ROI_1 - 12.5%
2) ROI_2 – 34.4%
3) ROI_3 – 20.5%

Probability (P_2) that women in selected group aged < 65 years have T – score < - 2.5 SD, if they have simultaneously Z – score > - 1.0 SD, is for:

1) ROI_1 - 0%
2) ROI_2 – 0%
3) ROI_3 – 0%

No woman was found in the female group aged < 65 years for measurements in all ROIs, who would have T - score < - 2.5 SD and, simultaneously, Z – score > - 1.0 SD.

Therefore the RR for fracture risk in the sample equals infinity, what means that the RR for women with T – score < - 2.5 SD and, simultaneously, Z – score ≤ - 1,0, is infinite times higher than for women with T – score < - 2.5 SD and, simultaneously, Z – score > - 1.0 SD.

RR in the population cannot be calculated, because RR in the sample = ∞.

We can only state that the higher fracture risk according to T – score is for women aged < 65 years, when there is a higher fracture risk also according to Z - score.

There was a statistical significance of association intensity (Table 4b, 5b, 6b) for dichotomous variables in contingency tables (Table 4a, 5a, 6a) for all ROIs. The closest association was found for the contingency table Table 5a (ROI_2), r = 0.546 (Table 5b).

Table 4a. Contingency table with categorized T and Z - scores for women aged < 65 and for women aged ≥ 65 (ROI_1)

Age < 65 years			
Z – score (SD) ROI_1 (count)	T – score (SD) ROI_1 (count)		Total count in row
	> - 2.5	≤ - 2.5	
> - 1.0	306 d	0 c	306
≤ - 1.0	14 b	2 a	16
Total count in column	320	2	322
Age ≥ 65 years			
> - 1.0	168 d	4 c	172
≤ - 1.0	1 b	3 a	4
Total count in column	169	7	176

Table 4b. Intensity of association for dichotomous variables according to age category for ROI_1 in Table 4a

Age < 65 years		
Coeficient of assotiation	Intensity of assotiation	Approximate significance
Phi	0.34573	*
Cramer´s V	0.34573	*
Age ≥ 65 years		
Phi	0.55422	*
Cramer´s V	0.55422	*

$\alpha = 0.05$ $p > 0.05$.

Table 5a. Contingency table with categorized T and Z - scores for women aged < 65 and for women aged ≥ 65 (ROI_2)

Age < 65 years			
Z – score (SD) ROI_2 (count)	T – score (SD) ROI_2 (count)		Total count in row
	T- score > - 2.5	≤ - 2..5	
Z - score > - 1.0	261 d	0 c	261
≤ - 1.0	40 b	21 a	61
Total count in column	301	21	322
Age ≥ 65 years			
> - 1.0	110 d	16 c	126
≤ - 1.0	1 b	49 a	50
Total count in column	111	65	176

Table 5b. Intensity of association for dichotomous variables according to age category for ROI_2 in Table 5a

Age < 65 years		
Coeficient of assotiation	Intensity of assotiation	Approximate significance
Phi	0.54636	*
Cramer´s V	0.54636	*
Age ≥ 65 years		
Phi	0.79709	*
Cramer´s V	0.79709	*

$\alpha = 0.05$ $p > 0.05$.

Table 6a. Contingency table with categorized T and Z – scores for women aged < 65 and for women aged ≥ 65 (ROI_3)

Age < 65 years			
Z – score (SD) ROI_3 (count)	T – score (SD) ROI_3 (count)		Total count in row
	T- score > - 2.5	≤ - 2.5	
Z - skóre > - 1.0	283 d	0 c	283
≤ - 1.0	31 b	8 a	39
Total count in column	314	8	322
Age ≥ 65 years			
> - 1.0	147 d	25 c	172
≤ - 1.0	0 b	4 a	4
Total count in column	147	29	176

Table 6b. Intensity of association for dichotomous variables according to age category for ROI_3 in Table 6a

Age < 65 years		
Coeficient of assotiation	Intensity of assotiation	Approximate significance
Phi	0.42997	*
Cramer´s V	0.42997	*
Age ≥ 65 years		
Phi	0.34334	*
Cramer´s V	0.34334	*

$\alpha = 0.05$ $p > 0.05$.

Probability (P) and Relative Risk (RR) for Women ≥ 65 Years

ROI_1

$$P_1 = \frac{a}{a+b} = \frac{3}{4} = 0.75 = 75\%$$

$$P_2 = \frac{c}{c+d} = \frac{4}{172} = 0.023 = 2.3\%$$

$$RR = \frac{0.75}{0.023} = 32.61$$

ROI₂

$$P_1 = \frac{a}{a+b} = \frac{49}{50} = 0.98 = 98\%$$

$$P_2 = \frac{c}{c+d} = \frac{16}{126} = 0.127 = 12.7\%$$

$$RR = \frac{0.98}{0.127} = 7.72$$

ROI₃

$$P_1 = \frac{a}{a+b} = \frac{4}{4} = 1.0 = 100\%$$

$$P_2 = \frac{c}{c+d} = \frac{25}{172} = 0.145 = 14.5\%$$

$$RR = \frac{1.0}{0.145} = 6.9$$

Probability (P_1) that women in selected group aged ≥ 65 years have T – score < - 2.5 SD, if they have simultaneously Z – score ≤ - 1.0 SD, is for:

1) ROI_1 - 75%
2) ROI_2 – 98%
3) ROI_3 – 100%

Probability (P_2) that women in selected group aged ≥ 65 years have T – score < - 2.5 SD, if they have simultaneously Z – score > - 1,0 SD, is for:

1) ROI_1 - 2.3%
2) ROI_2 – 12.7%
3) ROI_3 – 14.5%

Relative risk (RR) for sustaining fracture in the sample is for:

1) ROI_1 - 32.61 times
2) ROI_2 – 7.72 times
3) ROI_3 – 6.90 times

higher for women aged ≥ 65 years, having T – score < - 2.5 SD and, simultaneously, Z – score ≤ - 1,0 SD, than for the women having T – score < - 2.5 SD and, simultaneously, Z – score > - 1.0 SD.

In the population the RR is with probability 95% from the interval:

1) for ROI_1 – big scatter in the interval, because there are small counts for a, b, c, so the result is not valid
2) $\langle 1.09; 54.58 \rangle$ for ROI_2
3) for ROI_3 the RR cannot be calculated, because b equals 0 in the contingency table

There was a statistical significance of association intensity (Table 4b, 5b, 6b) for binary variables in contingency tables (Table 4a, 5a, 6a) for all ROIs. The closest association was found for the contingency table Table 5a (ROI_2), r = 0.797 (Table 5b)

Discussion

T and Z – score values are different in measurements by DXA densitometer in various skeleton sites not only in the same osteoporotic patient, but also in a healthy individual. This is conditional on the uneven bone mass loss in the skeleton.

In measurement sites with fast bone mass loss (DXA - vertebrae in lateral projection, DXA - Ward's area) the WHO criterion for T-score overestimates pathologies of BMD, and, conversely, in sites with a slow bone mass loss (DXA - total hip, DXA - trochanter) it underestimates pathologies of BMD [11, 12]. Intermediate bone mass loss is found when DXA is used at the distal forearm, at the vertebrae in AP projection and at the femoral neck area.

Therefore, the WHO set the BMD measurement by DXA instruments in the proximal femur area (total hip) as a gold standard for osteoporosis diagnosis. According to the WHO criteria, the osteoporotic fracture risk increases for both T and Z scores for every –1.0 standard deviation, and the increased fracture risk for the patient is imminent when the measurement results for T-score < - 2.5 SD and for Z-score ≤ - 1.0 SD. We used these criteria in statistical categorisation of both T and Z scores as the presence of increased fracture risk and the absence of increased risk.

In the study we determined how many times the fracture risk in women having both T and Z-score values within the increased risk category differs from the risk in women having only the T-score values within the increased fracture risk category. The study confirmed that women having both T and Z-scores within the increased fracture risk category face statistically significant higher fracture risk for all measured regions of interest in proximal femur area in comparison with women having only the T-score in the increased risk category. The highest RR of fracture for the whole sample of women according to T plus Z-score values was found in ROI_1(Neck). When the sample was divided into the old-age women (≥ 65 years) and the women aged < 65 years, we found out that in the latter group there was no woman having T-score < - 2.5 SD and, simultaneously, Z – score ≤ - 1.0 SD for measurements in all three ROIs. In the group of women under the age of 65 the RR of fracture equals infinity, there is an infinitely more times higher RR for women having T-score < - 2.5 SD and simultaneously, Z-score ≤ - 1.0 SD, than for women having T-score < - 2.5 SD and simultaneously, Z-score > - 1.0. This result is of high statistical significance, demonstrating that it is very important in the fracture risk assessment in postmenopausal women under the age of 65 to evaluate also the decrease of the number of standard deviations for Z-score. In the old age group there was a very high probability that if a woman had T-score < - 2.5 SD, she had also Z-score v ≤ - 1.0 SD. For ROI_3 (total hip) it reached even 100%! It means that for ROI_3 all women had Z-score ≤ - 1.0 SD, if they had T-score < - 2.5 SD. The fracture risk in old women (> 65 year) was the highest one for $ROI_{1,}$ as in the whole cohort, if both T and Z scores were within the increased fracture risk category.

The intensity of association of categorized T and Z scores was statistically significant for all selected relations. The highest intensity of association was found for the values measured in ROI_2 in the whole cohort, in the group of women under the age of 65 and also in the old women group (> 65 years), where it reached the value of 0.797. This highest intensity of association is conditional on the measurement of greater fraction of trabecular bone in the Ward's area in comparison with ROI_1 and ROI_3, and so measurement errors resulting from osteoarthrotic changes in Ward′s area are less significant. As Ward's area has a quicker bone loss than other measurement sites, it is not used for osteoporosis diagnosis with DXA instruments.

We have not come across a similar solution of the problem in the available literature, and, therefore, we do not have the possibility to compare our results with other authors. However, there are many studies comparing the sensitivity and specificity of T- and Z - scores for differentiating osteoporotic women from healthy ones by different measurement techniques and different measurement sites [13 – 18]. In these studies Z-score demonstrates a very good sensitivity and specificity. For example, in the study by Bergot C. et al., the authors proved that Z-score for QCT measurement had the highest sensitivity in differentiating osteoporotic women from the normal ones in comparison with T-score for QCT and with T - and Z- scores for DXA [19]. The precise BMD classification according to Z-score for osteoporosis diagnosis is not yet determined [20].

Conclusion

The study's contribution lies in the finding that the relative fracture risk in proximal femur area in women increases, if T-score < - 2.5 SD and, simultaneously, Z-score ≤ - 1.0 SD, and this risk is especially high for women under the age of 65.

In assessing the individual fracture risk in female osteoporotic patients (by questionnaires) we recommend to add also Z – score values from the measured area to T –score values from proximal femur area and to other anamnestic risk factors (age, sustained fracture after the age of 50, femur neck fracture in mother, increased biochemical markers of osteoresorbtion, smoking, body mass under 56 kg, and the like).

References

[1] Kanis JA, Johnell O, Oden A, Jonsson B, Dawson A, Dere W Risk of hip fracture derived from relative risk: An analysis applied to the population of Sweden. *Osteoporos. Int.* 2000; 11: 120–127.

[2] Kanis JA, Johnell O, Oden A, Oglesby AK, De Laet C, Jonsson B: Ten year probabilities of osteoporotic fracture according to BMD and diagnostic thresholds, *J. Bone Miner. Res.* 2001; 16 (Suppl 1): 194.

[3] Leslie WD, Metge C, Ward L :Contribution of clinical risk factors to bone density – based absolute fracture risk assessment in postmenopausal women. *Osteoporos. Int.* 2003; 14: 334 – 338.

[4] Black DM, Steinbuch L, Palermo L et al: An assessment tool for predicting fracture risk in postmenopausal women. *Osteoporosis Int.* 2001;12: 519 – 528.

[5] Leib ES, Lewiecki EM, Binkley N, Hamdy RC: Official positions of the international society for clinical densitometry. *J. Clin. Densitom.* 2004; 7: 1-6.

[6] Hilgers RD, Bauer P, Scheber V: Einfuhrung in die medizinische Statistik, Springer Verlag Berlin Heidelberg, 1. *Ausgabe*, 2003, 217 – 220.

[7] Pacakova V: *Štatistika pre ekonómov*, IURA EDITION, Bratislava, 1. vydanie, 2003, 193 – 234.

[8] Schumacher M, Schulgen Gabi : Methodik klinischer Studien. Verlag Berlin Heidelberg, 1. *Ausgabe*, 2002, 89 – 104.

[9] Matthis C, Weber T, O 'Neil W, Raspe H: Health impact associated with vertebral deformities: Results from the European Vertebral Osteoporosis Study (EVOS). *Osteoporos. Int.* 1998; 8: 364–372.

[10] Zhang J, Yu KF: What is relative risk? A method of correcting the odds ratio in cohort studies of common outcomes. *JAMA* 1998; 280: 1690–1691.

[11] Fualkner KG, von Stetten E, Steiger P, Miller P: Discrepancies in osteoporosis prevalence at different sceletal sites: impact on the WHO criteria. *Bone*1998; 23 (Suppl):194.

[12] Mazess RB: T – score differ among skeletal sites. *Calcif. Tissue Int.* 1999; 64: 104.

[13] Molyvda – Athanasopoulou E, Sioundas A, Kaltziioannou K: Dual X – ray absorptiometry reference data for Greek. *Europ. J. Radiol.* 2000; 36: 36–40.

[14] Henry MJ, Pasco JR, Pocock NA et al: Refence ranges for bone densitometers adopted Australia – wide: Geelong osteoporosis study. *Austral. Radiol.* 2004; 48:473- 475.

[15] McMahon K, Nightingale J, Pocock N: Discordance in DXA male reference ranges *J. Clin. Densitom.*, 2004;7:121–126.

[16] Wendlova J: Differences in distribution of T – scores and Z – scores among bone densitometry tests in postmenopusal women (a comparative study). *Wien. Med. Wschr.* 2002; 152: 591–595.

[17] Blake GM, Knapp KM, Fogelman I: Absolute fracture risk varies with bone densitometry technique used. A theoretical and in vivo study of fracture cases. *J. Clin. Densitom.* 2002; 5: 109–116.

[18] Newton LJ, Kenny RA, Frearson R, Francis MR: A prospective evalution of bone mineral density measurement in females who have fallen. *Age Ageing* 2003; 32:497–502.

[19] Bergot C, Laval – Jeantet AM, Hutchinson K et al: A comparison of spinal quantitative computed tomograpfy with dual energy X - ray absorbtiometry in European women with vertebral and nonvertebral fractures. *Calcif. Tissue Int.* 2001; 68:74–82.

[20] Unnanuntana A, Gladnick BP, Donnelly E et al: The assessment *J. Bone Joint Surg. Am.* 2010;92(3):743-53.

Chapter VI

Osteoporosis in a Female Population from Bratislava - Age-Related BMD Changes *(Epidemiological Study)*

Abstract

Patients and methods: We analysed 498 women (n = 498) in Bratislava (BA) population aged 21 to 90. We measured bone mineral density (BMD) in the area of proximal femur with one densitometric instrument DXA (dual energy X–ray absorptiometry), applying BMD and T-score values in three standard regions of interest: Neck (ROI_1), Ward ´s area (ROI_2), great trochanter (ROI_3).

Statistical analysis: 1. Chi-square goodness-of-fit statistics (χ^2 - test). 2. Analysis of variance (ANOVA).

Results: Measured values of T–score in ROI_1 , ROI_2 have normal distribution and a lognormal distribution of frequency in ROI_3. Using χ^2 - test (chi-square goodness-of-fit statistics), we determined the distribution of frequency of T-score values and the percentage of osteoporosis incidence in Bratislava female population. The osteoporosis incidence in Bratislava female population, according to T-score values measured in ROI_1 is 2.40%, in ROI_2 16.34% and in ROI_3 3.83% . Following the division of women into ten-year intervals, the statistically significant sample averages of T-score values were decreasing in relation to age only for ROI_2. The osteoporosis incidence in age intervals was rising with age for ROI_2, for ROI_1 the number of osteoporotic patients in the 61 to 70-year interval was lower than in the 41 to 50-year interval, and for ROI_3 the number of osteoporotic patients in the 51 to 60-year interval was lower than in the 41 to 50-year interval. Expect of the above mentioned intervals, T-score values decreased in relation to age also in ROI_1 and ROI_3. According to the analysis of variance, the age category explains 9.6% of the overall variability of T-score values for ROI_1, 24.7% for ROI_2 and 11.70% for ROI_3.

Conclusion: As in ROI_2 (Ward's area) greater fraction of the trabecular bone is measured in comparison with ROI_1 and ROI_3, the ROI_2 reflects best the age-related BMD changes. It is due to the fact that the trabecular bone is eight times more active metabolically than the cortical bone, and so it is more sensitive than the cortical bone for the determination of BMD changes. In ROI_1 and ROI_3 the relation was distorted by a greater fraction of cortical bone in comparison with ROI_2 and by an accidental selection of a higher number of osteoarthrotic patients in the 61 to 70-year interval for ROI_1 , and in the 51 to 60-year interval for ROI_3 . The age of patients is a significant risk factor. Bone mass decreases with age independently of the period of other risk factors activities, and also independently of the fact whether the patient suffers from primary or secondary osteoporosis (osteopenia). The development and application of densitometric methods measuring separately the trabecular and cortical bone excludes the distorting factor of osteoarthrosis and should be preferred in densitometric examination methods for a continuous observation of BMD changes.

Keywords: bone mineral density– proximal femur – osteoporosis – women – epidemiological study

INTRODUCTION

Skeletal osteoporosis is the reason for 90% of vertebrae and femoral neck fractures, 70% fractures in the distal forearm area, and about one third of Caucasian women over 50 sustain a fracture in one of these skeletal sites [1]. Therefore, two priorities are preferred in osteology: 1) active screening of patients with risk factors, who undergo densitometric examination; 2) reduction of individual fracture risk by complex treatment of osteopenia and osteoporosis [2,3].

The objective of densitometric measurement is the diagnosis of osteoporotic patients and in combination with anamnestic data and biochemical markers of osteoresorbtion, the determination of individual fracture risk. According to the WHO, a gold standard for osteoporosis diagnosis and the fracture risk assessment is DXA (dual-energy X-ray absorptiometry) in the proximal femur area (total hip). The problem of the DXA densitometry lies in the discrepancy of results given by densitometers from different manufacturers, in the discrepancy of BMD in measurements in various skeleton sites, differences in selected reference groups of healthy young adults, osteoarthrotic changes in the measured site, sclerotic plates in the aorta, scoliosis and obesity in patients [4, 5].

Objective of the Study

1. To determine the percentage of osteopenia and osteoporosis incidence in Bratislava female population.
2. To divide the sample group into ten-year intervals and to determine the number of osteoporosis cases (in %) in each age interval according to T-score values in the Neck (ROI_1), Ward's area (ROI_2) and great trochanter (ROI_3).
3. To determine which ROI detects best the age-related BMD changes.
4. To determine, using the analysis of variance, the percentage of age participation in the overall variability of T-score values for ROI_1, ROI_2 and ROI_3.

Characteristics of the Examined Patients Group and Methods Applied

General practitioners as well as specialists from different Bratislava districts were sending female patients with primary or secondary osteoporosis and osteopenia and patients with risk factors who did not undergo previous densitometric examination, to one densitometric centre. Women with risk factors were selected for densitometric examination in accordance with the indication list of the International Society for Clinical Densitometry (Official Positions 2004) [6].

So we had a random selection of patients with osteoporosis and osteopenia from Bratislava districts, and simultaneously, in the random selection of risk patients we diagnosed new patients with osteoporosis or osteopenia.

In the Bratislava population sample group there were 498 (n= 498) women of the average age of 59.71 ⟨58. 63; 60. 80⟩.

All women were examined with the same densitometer DXA (dual energy X–ray absorptiometry). Bone mineral density (BMD) was determined in standard regions of interest for DXA (ROI - region of interest) – Neck (ROI_1), Ward's area (ROI_2), great trochanter (ROI_3). BMD values were given in absolute numbers in *g* of Ca-hydroxyapathite crystals for cm^2 (g/cm^2), as well as in relative numbers as T-score (the number of standard deviations from the reference group of young healthy women) and Z-score (the number of standard deviations from the relevant age group of healthy women). Osteoporosis was diagnosed in accordance with the WHO criteria.

STATISTICAL ANALYSIS [6-8]

We have used the statistical program systems Statgraphics plus version 3 and SPSS version 6 for the analysis of the female group data.

1) Test for normality of distribution of T – score and Z - score values in ROI_1, ROI_2, ROI_3 (χ^2 – goodness-of-fit statistic, Kolmogorov – Smirnov statistics)
2) Division of the female cohort into 10–year age groups and determination of osteoporosis incidence percentage in each age group for T – score in ROI_1, ROI_2, ROI_3
3) Analysis of variance for comparison of mean T-score values for ROI_1, ROI_2, ROI_3 in age groups divided in ten-year intervals
4) Spearman rank correlation between variables of age and T-score for ROI_1, ROI_2, and ROI3.

RESULTS

Measured values of T – score in ROI_1 and ROI_2 have normal distribution (Graph 1, 2), lognormal distribution of probabilities is in ROI_3 (Graph 3).

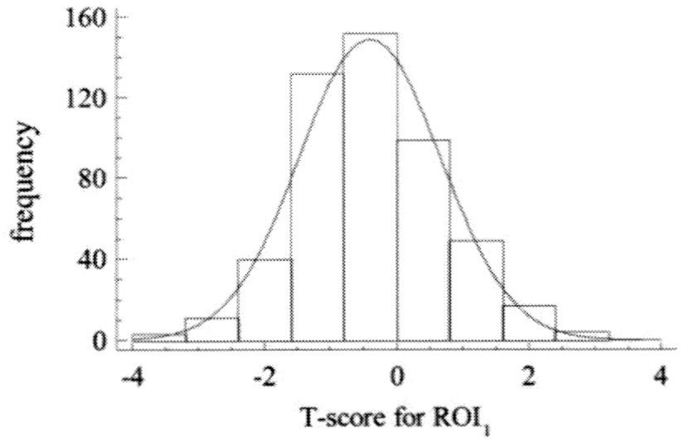

Graph 1. Normal distribution of T- score for ROI_1.

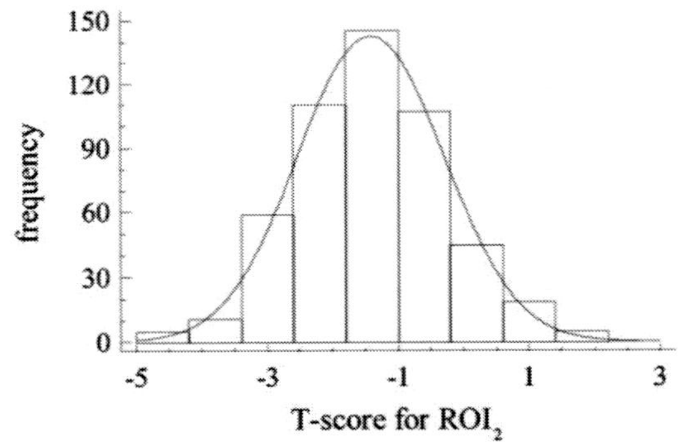

Graph 2. Normal distribution of T- score for ROI_2.

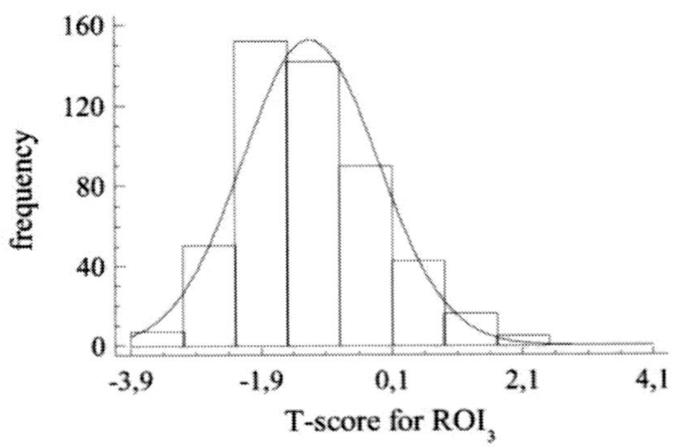

Graph 3. Lognormal distribution of T- score for ROI_3.

We determined the character of probability distribution by χ^2 – goodness-of-fit test for the comparison of empiric frequency distribution of T-score values found in the sample group with expected frequencies in the Bratislava female population at the significance level $\alpha = 0.05$ (tab.1). The percentage of expected number of women with osteoporosis and osteopenia in the Bratislava population is given in Table 1.

The percentage of osteoporotic patients in ten-year age intervals is given in Table 2.

Table 1. Observed and expected frequencies of osteoorotic female patients in Bratislava population (Chi – square goodness – of – fit statistic).

T – score (SD)	Observed frequency (count)	Expected frequency (count)	Expected frequency in %
ROI_1			
⟨ - 2.5; - ∞)	9	12	2.46
(- 2.5; - 1.0⟩	143	131	26.23
(- 1.0; 0.0⟩	181	179	35.8
(0.0;+ ∞)	165	176	35.3
ROI_2			
⟨ - 2.5; - ∞)	86	81	16.34
(- 2.5; - 1,0⟩	240	239	47.98
(- 1.0; 0.0⟩	123	127	25.39
(0.0;+ ∞)	49	51	10.29
ROI_3			
⟨ - 2.5; - ∞)	37	19	3.83
(- 2.5; - 1.0⟩	264	296	59.47
(- 1.0; 0.0⟩	130	120	23.99
(0.0;+ ∞)	67	63	12.71

$\alpha = 0.05$ $p > 0.05$.

χ^2 – square test for ROI_1 = 9.338 with 4 d.f. p - value = 0.053

χ^2 – square test for ROI_2 = 2.878 with 6 d.f. p – value = 0.824

χ^2 – square test for ROI_3 = 13.609 with 7 d.f. p – value = 0.058

1) ROI_1 – the percentage of osteoporotic patients is higher in the 51 to 60 age interval than in the 61 to 70 interval, otherwise the osteoporosis incidence percentage rises with age;
2) ROI_2 – osteoporosis incidence percentage rises with age in all age intervals;
3) ROI_3 – the percentage of osteoporotic patients is higher in the 41 to 50 age interval than in the 51 to 60 interval, otherwise the osteoporosis incidence percentage rises with age in all intervals.

Table 2. The sample: Count of female osteoporotic patients (in %) in 10 – year age groups for ROI_1, ROI_2, ROI_3

Age group	ROI 1		ROI_2		ROI_3	
	T score > - 2.5 SD n_1 n_2 (%)	T - score < - 2.5 SD n_1 n_2(%)	T score > - 2.5 SD n_1 n_2(%)	T - score < - 2.5 SD n_1 n_2(%)	T score > - 2.5 SD n_1 n_2(%)	T - score < - 2.5 SD n_1 n_2(%)
21 – 30	16 100.0	0 0	16 100.0	0 0	16 100.0	0 0
31 - 40	11 100.0	0 0	11 100.0	0 0	11 100.0	0 0
41 – 50	75 100.0	0 0	72 96.0	3 4.0	72 96.0	3 4.0
51 - 60	158 98.7	2 1,3	153 95.6	7 4.4	157 98.1	3 1.9
61 – 70	136 99.3	1 0.7	108 78.8	29 21.2	129 94.2	8 5.8
71 – 80	78 95.1	4 4.9	44 53.7	38 46.3	65 79.3	17 20.7
81 - 91	15 88.2	2 11.8	8 47.1	9 52.9	11 64.7	6 35.3

n_1 – count of female patients in age group.
n_2 - % of female osteoporotic (or nonosteoporotic) patients in age group.

Sample averages of T-score values in ROI_1, ROI_2, ROI_3 and their confidence intervals (C.I.) for set age groups are illustrated in Graphs 4, 5 and 6. They show that the values of sample intervals are decreasing with age. Statistically significant differences among compared T-score values from all set ten-year age intervals can be found only in ROI_2 (Graph 5).

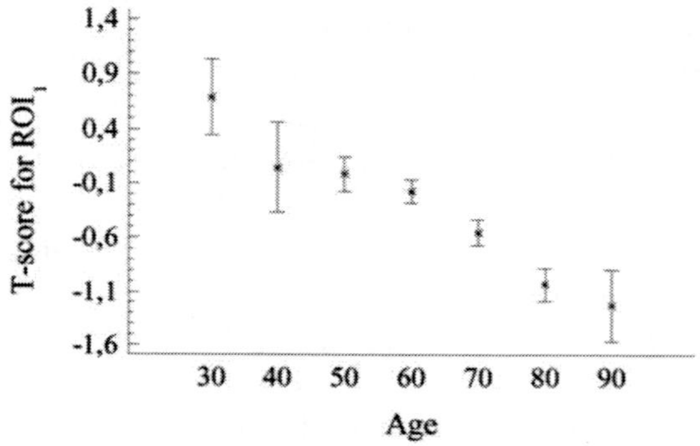

Graph 4. Averages and 95% confidence intervals for T – score in ROI_1.

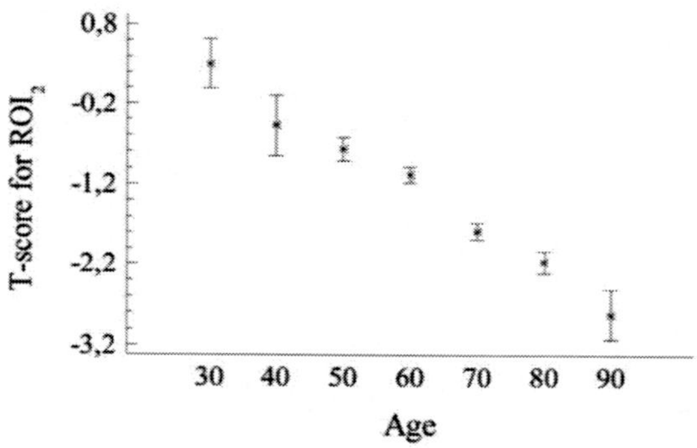

Graph 5. Averages and 95% confidence intervals for T – score in ROI_2.

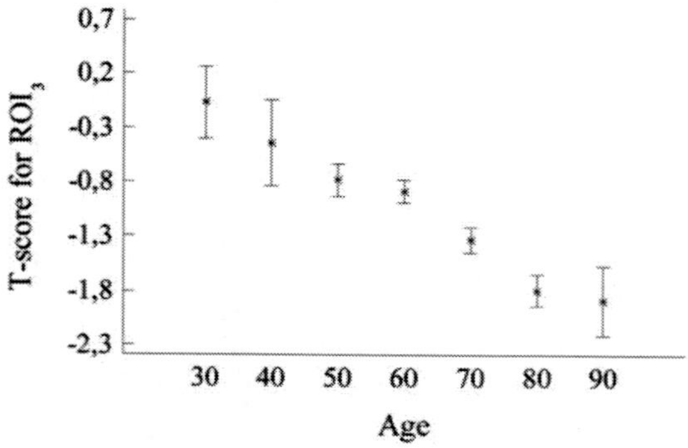

Graph 6. Averages and 95% confidence intervals for T – score in ROI_3.

In the variance analysis (ANOVA) the age category explains 9.6% of the overall variability of T-score values for ROI_1, 24.7% for ROI_2, and 11.7% for ROI_3 (Table 3).

Fisher's test (F–test) at the significance level of $\alpha = 0.05$ confirmed the significance of the age factor.

Statistically most significant negative correlation between age and T-score was determined for ROI_2 ($r = - 0.5785$).

Table 3. ANOVA analysis for age and T – score in ROI_1, ROI_2, ROI_3. P - value < 0.05 - alternative hypothesis H_1 is accepted

ROI_1	Sum of squares	F – ratio	p - value	Variance determined with age in %
Variance between groups	54.285	8.66	0.001	9.6
Variance within groups	512.912			
Total variance	567.197			
ROI_2				
Variance between groups	151.989	26.82	0.001	24.7
Variance within groups	463.761			
Total variance	615.75			
ROI_3				
Variance between groups	63. 227	10.85	0.001	11.7
Variance within groups	477.092			
Total variance	540.319			

F – ratio (Fisher – ratio).

Discussion

Osteoporosis incidence in Bratislava female population, according to T-score values measured in ROI_1 is 2.40%, in ROI_2 16.34% and in ROI3 is 3.83%. In comparison with the results of the European Vertebral Osteoporosis Study (EVOS) [9], which found the osteoporosis incidence in European women to be 12% according to BMD in vertebrae, the osteoporosis incidence in Bratislava female population is, according to measurements in proximal femur for ROI_1 and ROI_3 lower and for ROI_2 higher than European average values in EVOS study. Women aged 50 to 79 from 19 European countries were randomly selected for the EVOS study and were divided into groups according to 5-year age intervals. The age was the only criterion for the inclusion of women in the study [10]. The incidence of osteoporosis according to BMD values in vertebrae is higher than that according to BMD in neck or total hip in postmenopausal women [11–13]. In dividing the cohort of examined women into ten-year age intervals we found that the osteoporosis incidence percentage rises in each consequent age group for T-score values only in ROI_2 (Ward's) as well as the ROI is not acceptable for diagnosis of osteoporosis. As in ROI_1 and ROI_3 greater fraction of the cortical bone is measured in comparison with the ROI_2, the higher incidence of osteoporotic patients in the lower age interval than in the higher age interval is conditional on random selection of higher number of patients with arthrotic changes in measurement sited ROI_1 and $ROI_{3.}$ Through random selection a higher number of patients with osteoarthrosis in measured ROI_1 site got into the 61 to 70 age interval as well as in the 51 to 60 age interval for $ROI_{3.}$ Measurement in ROI_2 area, independently of osteoarthrosis, detects well the bone mass loss, even though the bone mass loss is faster in this area than in other parts of proximal femur [14, 15]; therefore, it is not used for osteoporosis diagnosis. Important finding is that Fisher's test confirmed the significance of age risk factor. Bone mass was decreasing with age, independently of other risk factors and time of their activities (which we have no evaluated), as well as independently of whether the patient had primary or secondary osteoporosis (osteopenia). The age-related decrease of BMD was reflected best in Ward's area. T-score values in this area showed the closest negative correlation with age ($r = 0.5785$). The trabecular bone is eight times more active metabolically than the cortical bone and, therefore, the methods measuring changes in the trabecular bone BMD are the most sensitive [5].

Exclusion of distorting factor of osteoarthrosis is supported by the development and application of quantitative computer tomography (QCT) densitometric method. QCT method measures separately the cortical and trabecular bones in the area of the vertebrae and proximal femur and gives the number of Ca-hydroxyapathite crystals in cubic measure units (g/cm^3), in contrast to DXA, which gives the number of Ca-hydroxyapathite crystals in square measure units (g/cm^2), and measures the cortical and trabecular bones together [5].

It follows from this that DXA and QCT do not provide identical information on BMD of measured bones. QCT measurements values are age-related, but independent of arthrotic calcification in the vertebrae and proximal femur areas, sclerotic plates in aorta, and patient's scoliosis. Therefore, the QCT examination is suitable for all age and weight categories [16]. Several authors [16–18] compared the sensitivity of DXA and QCT in differentiating a normal from an osteoporotic population. Higher sensitivity was found in QCT measurements, using t-test, logistic regression analysis, odds ration, ROC (receiver operator characteristics), and Wilcoxon's pair test. Highest sensitivity of differentiation of osteoporotic patients from healthy individuals was achieved by Z-score for QCT, in comparison: a) with T-score for QCT b) with T-score for DXA c) with Z-score for DXA. The introduction of a localiser into a scanner for QCT methods, as well as the determination of *mAs* dose for patients according to their height and weight, reduce the radiation load by 50% and more, and QCT methods acquire an important position in densitometry [19, 20].

Standard application of QCT methods requires a whole range of prospective clinical studies determining a relative fracture risk in a time interval in treated and non-treated postmenopausal women. The WHO criteria for BMD classification by means of T-score for DXA may not provide the best predictive ability to detect individuals with the greatest risk for fracture. They are novel approaches in DXA and QCT measurement methods, with the possibility to calculate the geometric values of proximal femur. These geometric values of proximal femur combined with BMD of proximal femur provide valuable clinical information, which is more sensitive for QCT, because here is the possibility to win 3D scan and to measure separately the cortical and trabecular bones in the selected area of the vertebrae and proximal femur.

Conclusions

Summing up

1. Osteoporosis incidence in Bratislava female population, according to T-score values measured in ROI_1 is 2.40%, in ROI_2 16.34% and in ROI_3 3.83%.
2. Patient's age is a significant risk factor for osteoporosis. Bone mass decreases with age, independently of the period of other risk factors activities, and also independently of the fact whether the patient suffers from primary or secondary osteoporosis (osteopenia).
3. Age category provides for the highest percentage in the variability of T-score values (24.7%) for measurements in ROI_2.
4. In all three measured ROIs, the statistically most significant negative correlation can be found between age and T-score measured in the trabecular bone in ROI_2 (Ward's area).
5. Separately measurement of the trabecular bone, which is eight times more active metabolically than the cortical bone, should be preferred in densitometric examination methods for a continuous observation of BMD changes.

References

[1] Melton LJ, Thamer M, Ray NF et al: Fractures attributable to osteoporosis: Report from the National Osteoporosis Foundation. *J. Bone Miner. Res.* 1997;12:16–23.

[2] Melton LJ, Chrisschilles EA, Cooper C et al: How many women have osteoporosis? *J. Bone Miner. Res.* 1992;7:1005–1110.

[3] Blake GM, Knapp KM, Fogelman I: Absolute fracture risk varies with bone density techniques used. A theoretical and in vivo study of fractures cases. *J. Clin. Densitom.* 2002; 5:109–116.

[4] McMahon K, Nightingale J, Pocock : Discordance in DXA male reference ranges *J. Clin. Densitom.* 2004;7:121–126.

[5] Liu G, Peacock M, Eliam O, et. al: Effect of osteoarthritis in the lumbar spine and hip on bone mineral density and diagnosis of osteoporosis in elderly men and women. *Osteoporos. Int.* 1997;7:564-569.

[6] Leib ES, Lewiecki EM, Binkley N, Hamdy RC: Official positions of the international society for clinical densitometry *J. Clin. Densitom.* 2004; 7:1-6.

[7] Hilgers RD, Bauer P, Scheber V: Einfuhrung in die medizinische Statistik Springer Verlag Berlin Heidelberg, 1. *Ausgabe*, 2003, 217–220.

[8] Pacakova V: *Štatistika pre ekonómov* (Statistics for economists), IURA EDITION, Bratislava, 1st edition, 2003, 193–234.

[9] Matthis C, Weber T, O 'Neil W, Raspe H: Health impact associated with vertebral deformities: Results from the European Vertebral Osteoporosis Study (EVOS) *Osteoporos. Int.* 1998; 8:364–372.

[10] Bartl R, Bartl Ch: Osteoporose – Manual. Diagnostik, Prävention und Therapie, Springer Verlag Berlin, 1. *Ausgabe*, 2004, 201.

[11] Kanis JA, Gluer CC: An update on the diagnosis an assessment of osteoporosis with densitometry. *Osteopor. Int.* 2000; 11:192–202.

[12] Oden A, Dawson A, Dere W et al: Lifetime risk of hip fractures is underestimated *Osteoporos. In.* 1998; 8: 599–603.

[13] Kanis JA, Melton LJ, Christiansen C et al: The diagnosis of osteoporosis *J. Bone Miner. Res.* 1994; 9:1137–1141.

[14] Faulkner KG, von Stetten E, Steiger P, MIller P: Discrepancies in osteoporosis prevalence at different skeletal sites: impact on the WHO criteria *Bone* 1998; 23(Suppl):194.

[15] Mazess RB: T – score differ among skeletal sites. *Calcif. Tissue Int.* 1999; 64: 104.

[16] Bergot C, Laval – Jeantet AM, Hutchinson K et al: A comparison of spinal quantitative computed tomography with dual energy X-ray absorptiometry in European women with vertebral and nonvertebral fractures. *Calcif. Tissue Int.* 2001; 68: 74–82.

[17] Crabtree N, Ward K: Bone Densitometry: Current Status and Future Perspectives. *Endocr. Dev.* 2009;16:58-72.

[18] Beck TJ: Comparison of DXA hip structural analysis with volumetric QCT. *J. Clin. Densitom.* 2009;12(2):260-1.

[19] Adams JE: Quantitative computed tomography. *Eur. J. Radiol.* 2009;71(3):415-24.

[20] Khoo BC, Brown K, Cann C et al: Comparison of QCT derived and DXA derived Areal bone minerqal density and T – scores. *Osteoporos. Int.* 2009;20(9):1539-45.

Chapter VII

Chondrosis of the Disc – Risk Factor for Osteoporotic Vertebral Fractures *(Biomechanical Analysis)*

Abstract

Based on biomechanical analysis, we present an until–now unrecognised new view on pathological interactive relations in basic functional motor segments of the spine (vertebra – disc – vertebra), elevating the risk of osteoporotic vertebral fractures. They are classified as follows: 1. Degenerative alterations of intervertebral disc (chondrosis); 1a) *decrease of intervertebral disc viscoelasticity;* 1b) *increase of compressive and tensile stress in the intervertebral disc.* Loading the spine increases the compressive and tensile stress in the disc, afflicted by chondrosis, beyond physiological values. This increase conditions the non-physiological elevation of reaction compression forces in adjacent vertebrae, representing a fracture risk for endplates of these vertebrae. 2. Osteoporosis of vertebrae 2a) *decrease of vertebra elasticity and strength;* 2b) *phenomenon of local elevated compressive stress in the vertebra fracture site.* In the vertebral body endplate infraction or fracture site develop a locus minoris of resistance and it contributes to the progression of breaking the vertebra and its whole compression by loading the spine with compressive force. With regard to the fact that abovementioned risk factors are influenceable, we suggest preventive measures.

Keywords: chondrosis – osteoporosis – viscoelasticity – hysteresis loop – elasticity and strength- vertebral fractures

INTRODUCTION

In osteological clinical practice there are many recognized risk factors for the development of osteoporosis and its complications – fractures. Rapid advances in scientific research reveal new risk factors again and again. They can be divided into infuenceable and non-influenceable factors. For the influenceable risk factors there are preventive and therapeutic measures which eliminate or reduce their influence (exposure) upon the patient aiming at the prevention of the development of osteoporosis and its complications.

Based on the results of biomechanical analysis we present in this paper some information about an until-now-unrecognised risk factor for vertebral fracture, namely chondrosis, a degenerative disease of intervertebral discs. We regard as a positive fact that this factor can be classified as influenceable. The development of chondrosis may be prevented by a proper lifestyle from young age: 1. rational diet and maintaining normal BMI values; 2. regular physical activity targeted at keeping the spine flexible; 3. ergonometry; 4. elimination of non-physiological biomechanical overloading of the spine; 5. observation of a proper daily drinking regimen; 6. elimination of bad habits: consumption of alcohol, black coffee, smoking. To achieve and maintain a functional motor performance of the spine it is necessary to support the physiological viscoelasticity of discs and the physiological elasticity of vertebrae by a proper regimen.

BIOMECHANICAL ANALYSIS

Viscoelasticity is a mechanical property of biomaterial. In the mechanics of solids there are plastic, elastic, and viscoelastic characteristics of materials.

Tissues and organs in human organism are prevailingly viscoelastic or sporadically elastic. Therefore, we are going to deal with these two notions only. We hold it important that physicians are aware of the biomechanical principles of elasticity and viscoelasticity. This knowledge helps them understand the biomechanics of loading the physiological spine and the spine afflicted by osteoporosis and chondrosis, resulting in better diagnostic and therapeutic approach by the physicians towards their patients.

THE DIFFERENCE BETWEEN ELASTIC AND VISCOELASTIC BIOMATERIAL

Elasticity [1]

After removing the application of deformation force, whose magnitude is smaller than the elastic limit of biomaterial, the biomaterial returns to its original state. This is the characteristic property of elastic biomaterial.

Elastic biomaterial stores 100% of the applied deformation force (energy) for its deformation. The stress - strain curve with increasing loading by deformation force (compression) as well as the stress - strain curve with decreasing unloading by deformation force are both identical, i.e., the elastic biomaterial changes its form in the same way by loading and unloading. The linear part of the stress-strain curve holds true to Hook's law, and with this law it is possible to calculate elasticity modulus for relevant biomaterial (in this case bone – vertebra) (Figure1).

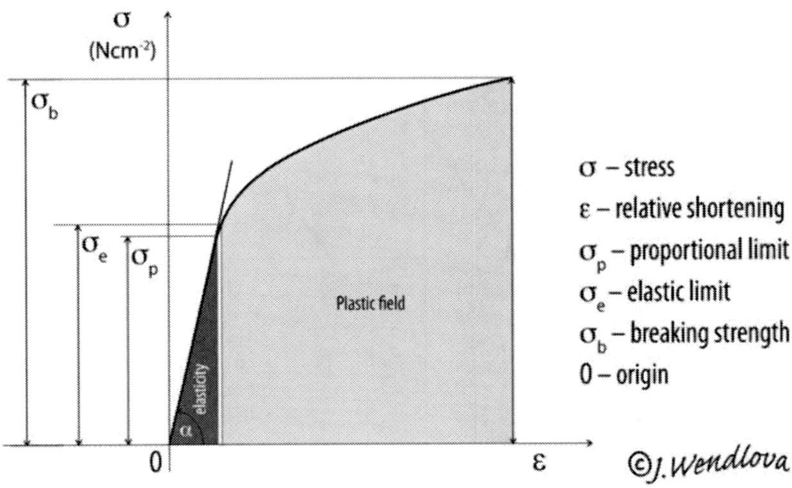

Figure 1.

Viscoelasticity [2]

Viscoelasticity is the property of biomaterials that exhibit both viscous and elastic characteristics. After the application of deformation force, viscoelastic material does not return to its original form and does not store 100% of the applied deformation force (energy) to its deformation, but a portion of this energy is dissipated (lost) in the vicinity of the biomaterial. Viscoelasticity with a characteristic dissipation of energy is represented by a hysteresis loop. Hysteresis loop is a typical curve of dependence between stress and strain for viscoelastic materials. The stress- strain curve with increasing loading by deformation force and the stress - strain curve with decreasing unloading by deformation force are not identical. This means that viscoelastic biomaterial acquires another form (strain) by being loaded with increasing deformation force as well as another form (strain) by being unloaded with decreasing deformation force. The inner area of the hysteresis loop is equal to the energy lost in the loading cycle (Figure 2).

Viscoelastic materials have three typical properties:

1. dissipation (loss) of a portion of deformation energy
2. stress relaxation
3. creep

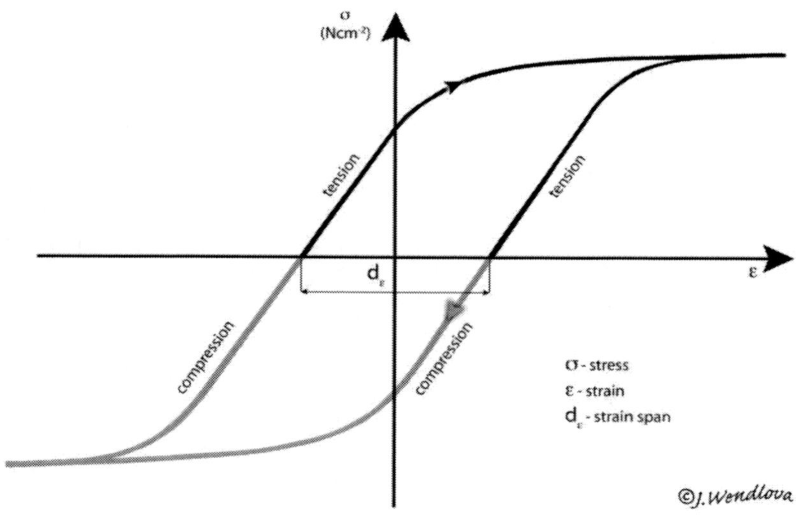

Figure 2. Stress-strain curve for viscoelastic materials. Cyclically loading in tension and compression of vertebral endplates (hysteresis loop).

Stress Relaxation [2]

Viscoelastic material exhibits maximum stress and maximum strain by application of maximum permitted deformation force, which does not cause the disruption of biomaterial. After the end of deformation force application, the stress within the material decreases with time up to the limiting stress, with persisting maximum strain. This phenomenon is called stress relaxation in viscoelastic biomaterial (Figure 3).

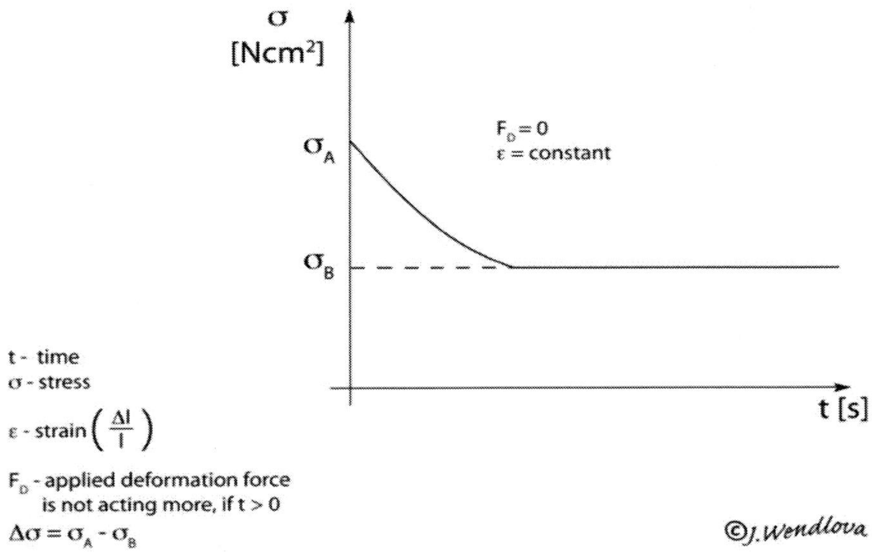

Figure 3. Stress relaxation.

Creep [2]

Essentially, creep is the opposite of stress relaxation. If the magnitude of deformation force is held constant (by constant stress in biomaterial), the strain (deformation) of the biomaterial increases until reaching maximum strain, persisting for a certain time. Creep phenomenon requires the condition that the deformation force magnitude is smaller than the magnitude of critical deformation force, conditioning the disruption of viscoelastic biomaterial (Figure 4).

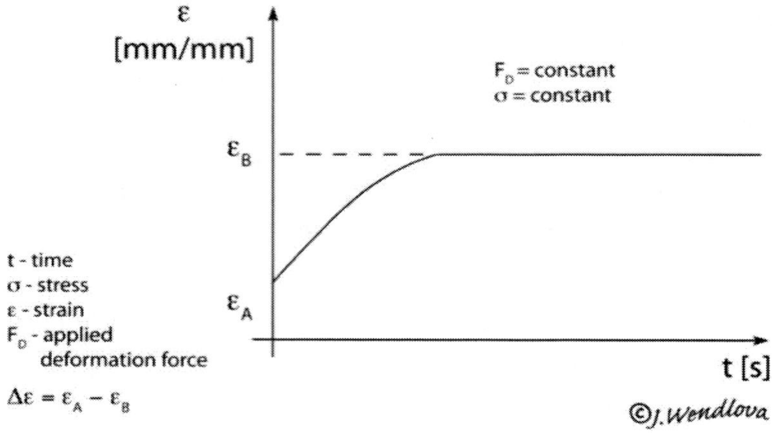

Figure 4. Creep.

INTERVERTEBRAL DISC

Intervertebral disc is a typical biomaterial, namely fibrocartilage, meeting the criteria for viscoelasticity. It consists of three anatomic and biomechanical components:

1. nucleus pulposus (NP) – central jellylike nucleus
2. annulus fibrosus (AF) – concentric layers of collagen fibres around nucleus pulposus
3. viscous fluid , a component of both nucleus pulposus and annulus fibrosus

Annulus fibrosus contains more viscous fluid than nucleus pulposus, therefore, nucleus pulposus has more characteristics of elastic biomaterial, while annulus fibrosus has those of viscoelastic material.

BIOMECHANICAL ANALYSIS OF THE FUNCTION OF A HEALTHY INTERVERTEBRAL DISC [3-6]

In Figure 5 we simulate the loading of the basic functional motor segment (vertebra – disc – vertebra) by vertical compression force. Compressive stress distribution is equal in the disc. A part of the compression force is transferred from nucleus pulposus to the brim of annulus fibrosus as tension force. This tension force is transferred to the disc edge, to a 1 to 4 mm wide circular zone, and conditions the bulging of the disc along the whole circumference of the disc (the disc bulges outwards). The bulged disc overlaps the edges of two adjacent vertebra bodies. Healthy disc is eccentrically bulged in the range up to 1.0 mm. Eccentrically located at the disc edge, fibrous rings in annulus fibrosus are more densely arranged near the nucleus pulposus, to better resist the elevated stress in tension. The angled, lamellar structure of the annulus fibrosus is integral to its load – bearing function. Reorientation of this fibre structure with applied load may contribute to nonlinear mechanical behaviour and to large increases in tensile modulus [7].

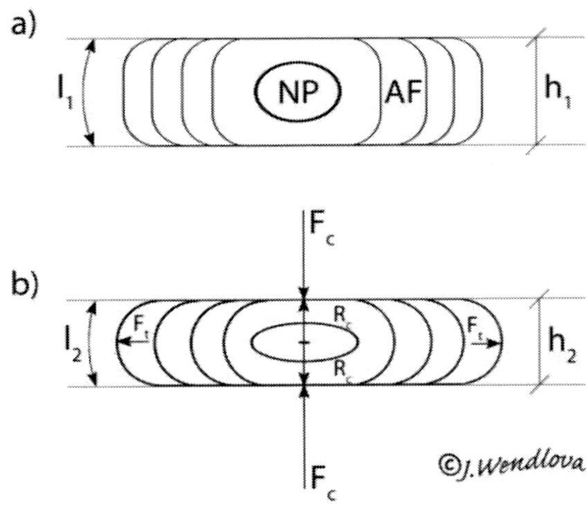

Figure 5. Model of intervertebral disk. a) without load. b) with load.

A part of the compression energy applied to the disc dissipates into its surroundings (energy loss) due to its viscoelasticity – the disc diminishes the impact of the compression force. Compressive stress, developing in the disc during its loading, is transferred as the compression force applied to the area of adjacent vertebra bodies. Reaction compression forces develop in the vertebrae (reaction compressive stress), and apply against compression forces in the disc. During flexion and extension of the spine, disc viscoelasticity allows an equal distribution of reaction compressive stress in adjacent vertebrae. This is very important for the equal stress of vertebrae (Figure 6).

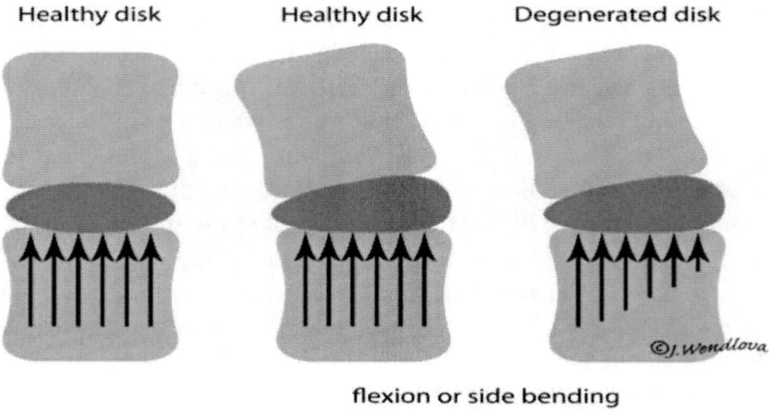

Figure 6. The distribution of the compressive stress over the vertebral endplates.

Biomechanical Analysis of the Function of a Disc Afflicted by a Degenerative Process

Degenerative process of intervertebral discs is manifested primarily by the decrease of their viscoelasticity, i.e., by alteration of the disc quality. The disc acquires more of the properties of an elastic than viscoelastic biomaterial, with undesired advancing decrease of the elasticity modulus, i.e., with little ability to resist deformation changes without damaging the disc. During flexion and extension of the spine there is no desired dissipation of part of the deformation energy applied to the disc, due to the decrease or loss of the viscoelasticity of the intervertebral disc. There is a decrease of the disc ability to dampen the effects of compression forces applied to the disc. During loading of the spine, annulus fibrosus is not able of circular and eccentric bulging from intervertebral spaces, and it results the increase of compressive stress (tensile stress at the disc edge) in those disc sites with greater pressure [8- 10]. Pathological increase of compressive stress in nucleus pulposus and annulus fibrosus is transferred as an increasing compression force applied to the area of endplates of adjacent vertebrae. In the vertebra endplates there occurs unequal distribution of reaction compressive stress, increasing ventrally by flexion, dorsally by extension, laterally by side bends, and reaches much higher values as during stress of a healthy disc. This disproportional increase of compressive stress in the ventral, dorsal or lateral part of vertebrae in dependence of load direction significantly elevates the risk of a fracture or infraction of endplates of osteoporotic vertebrae [11, 12] (Figure 6).

Biomechanical Analysis of the Function of the Healthy Vertebra

The vertebra, as a bone, is classified among solid elastic biomaterials with a typical diagram of strain dependence upon stress, developed during loading of the vertebrae by compression or tension (Fig 1). Elasticity modulus of the cortical part of the vertebra is many times higher than the elasticity modulus of the trabecular part of the vertebra. This elasticity of the cortical part is desired with regard to the contact of the cortical part of the vertebra with a viscoelastic disc [13-15]. While loading the spine by vertical compression force, the compressed disc is pressed into the vertebra body due to the elasticity of the vertebra body endplate approximately up to 0.5 mm [4], with maximum bulging in the centre of the endplate (endplate bulges inwards).

Biomechanical Analysis of the Function of the Osteoporotic Vertebra

As a consequence of osteoporosis, the elasticity and strength of cortical and trabecular vertebra parts are reduced. While loading the spine, fractures and infractions may occur in the cortical part in the endplates, and, consequently, also in the trabecular part of the vertebra[16].

In the site of an infraction or fracture of the endplate, a defect occurs, which is very adverse with regard to biomechanical stress. During the vertebra exertion by loading the spine with vertical compression forces, the compressive stress in the defect site is several times higher in comparison with the compressive stress at undamaged sites of the endplate. This phenomenon is termed a *phenomenon of increased local stress* in biomechanics (Figure7) and represents locus minoris of resistence. While loading the spine, due to the increased compressive stress, the magnitude of the defect in the endplate gradually increases, until it leads to the development of compression of the whole vertebra body [17].

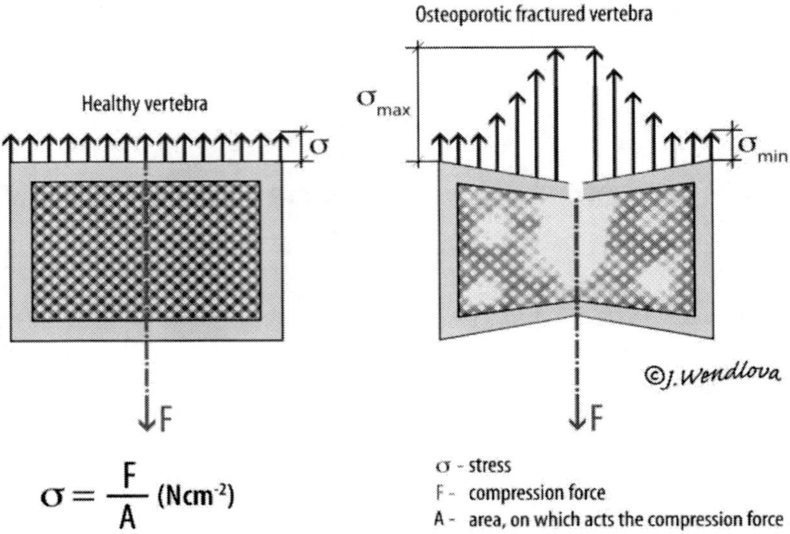

Figure 7.

Conclusion

The finding that the degenerative disease of the disc is a new risk factor for fractures of vertebra endplates in osteoporotic patients should lead to the following preventive measures:

1. we suggest additional examinations of each osteoporotic patient, namely:

   a) functional examination of the spine
   b) detailed assessment of the X-ray of the spine in lateral and anterioposterior projection – it is important to notice not only the height (frontal, central, posterior) and deformations of the vertebrae, but to assess also the alteration on discs and to determine vertebrae at high risk of endplate fractures according to the determination of discs afflicted by chondrosis

2. education of patients – observation of drinking regimen to support the viscoelasticity of the disc; (calculation of required intake of fluids

according to the weight of the patient and season); rational diet with emphasis on a sufficient supply of nutrients needed for the regeneration of connective tissue, cartilage and bones; maintaining normal BMI values; kinesitherapy aimed at supporting the flexibility of the spine (preferring exercises in a swimming pool, and in a supine position); ergonometry; smoking and alcohol consumption ban.

3. patients with chondrosis and osteoporosis to be included in a long-term rehabilitation programme; exercises in patient groups under professional supervision.

RESUME

Pathological interactive relation in basic functional motor segments of the spine, elevating the risk of osteoporotic vertebral fractures, can be summed up as follows:

1. degenerative alterations of intervertebral disc (chondrosis):

   - decrease of intervertebral disc viscoelasticity;
   - increase of compressive and tensile stress over physiological values in the degenerative disc during its loading.
   - loss of the disc's ability to dissipate a part of the deformation energy applied to the disc, i.e., loss of the ability to dampen the effects of deformation forces, namely shock deformation forces (e.g., jumping, swinging movements)

2. osteoporosis of vertebrae:

   - decrease of vertebral elasticity and strength;
   - phenomenon of local elevated compressive stress in the vertebra fracture site -a locus minoris of resistance develops in the vertebral body endplate infraction or fracture site, and it contributes to the progression of breaking the vertebra and its whole compression.

For the prevention of chondrosis and osteoporosis it is important to educate families about a proper lifestyle and active approach of individuals to health issues.

For the therapy of patients with chondrosis and osteoporosis it is also important to educate patients about a proper regimen, to motivate them to change to the desired lifestyle and to include them in a long-term rehabilitation programme.

Declaration – Conflict of Interests

The author declares that she has no competing interests (financial or non financial).

References

[1] Wendlová J. Bone Quality. Elasticity and Strength. *Bratisl. lek. listy,* 2008;109:383—6.

[2] Lakes R. *Viskoelastic solids*,1 st ed. Cambridge: University Press; 2009. pp. 36 – 39.

[3] Adams MA, Mc Nally DS, Dolan P. Stress disturbations inside intervertebral disc. *J. Bone Jt. Surg*., 1996; 78B: 965–972.

[4] Zhao FD, Pollintine P, Hole BD et al: Vertebral fractures usually affect the cranial endplate because it is thinner and supported by less-dense trabecular bone. *Bone*. 2009;44(2):372-9.

[5] Cannella M, Arthur A, Allen S et al. The role of the nucleus pulposus in neutral zone human lumbar intervertebral disc mechanics. *J. Biomech.,* 2008;41:2104-11.

[6] Horst M, Brinckmann P. Measurement of the distribution of axial stress on the endplate of the vertebral body. *Spine*, 1981; 6: 21 – 32.

[7] Guerin HA, Elliott DM. Degeneration affects the fiber reorientation of human annulus fibrosus under tensile load. *J. Biomech*., 2006;39:1410-8.

[8] Colombini A, Lombardi G, Corsi MM, et al. Pathophysiology of the human *Int. J. Biochem. Cell Biol*., 2008;40:837-42.

[9] Rohlmann A, Zander T, Schmidt H et al. Analysis of the influence of disc degeneration on the mechanical behaviour of a lumbar motion segment using the finite element method *J. Biomech*.,2006;39:2484-90.

[10] Kim PK, Branch CL Jr. The lumbar degenerative disc: confusion, mechanics, management *Clin. Neurosurg*., 2006;53:18-25.

[11] Rohlmann A, Zander T, Schmidt H et al. Analysis of the influence of disc degeneration on the mechanical behaviour of a lumbar motion segment using the finite element method *J. Biomech.,* 2006;39:2484-90.

[12] Tsantrizos A, Ito K, Aebi M et al. Internal strains in healthy and degenerated lumbar intervertebral discs *Spine*, 2005;30:2129-37.

[13] Bevill G, Farhamand F, Keaveny TM: Heterogeneity of yield strain in low-density versus high-density human trabecular bone. *J. Biomech.* 2009;42(13):2165-70.

[14] Anup S, Sivakumar SM, Suraishkumar GK.: Influence of viscoelasticity of protein on the toughness of bone *J. Mech. Behav. Biomed. Mater.* 2010;3(3):260-7.

[15] Homminga J, Mccreadie BR, Weinans H et al. The dependence of the elastic properties of osteoporotic cancellous bone on volume fraction and fabric. *J. of Biomech.*, 2003; 36:1461-67.

[16] Einhorn TA. The structural properties of normal and osteoporotic bone. *Instruct. Course Lect.*, 2003;52:533–539.

[17] Wendlová J: Osteoporoza. Pohybova liečba (Osteoporosis. Kinesitherapy). 1st ed, Bratislava, *Sanoma Magazines Slovakia*, 2008. pp. 50-2.

Chapter VIII

The Biomechanical Methods Used in Determining Bone Quality Elasticity and Strength *(Biomechanical Analysis)*

Abstract

The aim of the article is to explain in more detail the biomechanical methods used in determining bone quality as well as to describe basic characteristic bone qualities resulting in the application of these methods. Mechanical properties of biomaterials are characterised by stress – strain curves, produced on the basis of testing the biomaterials by progressive tensile and compressive load. According to these curves, the bone counts among brittle materials. Stress – strain curves under tensile and compressive load for brittle materials comprise proportional limit, elastic limit, and breaking strength. Recognition of biomechanical characteristics of bone quality (proportional limit, elastic limit, breaking strength) helps physicians to understand the importance of measuring the new parameters in bone densitometry by DXA (dual energy X-ray absorptiometry) such as femur strength index (FSI), and safety factor (SF).

Keywords: bone quality – biomechanics – proportional limit – elastic limit – breaking strength - compressive load – tensile load

INTRODUCTION

The quality of osteoporotic bone, elasticity and strength of bone – these are the terms often used in professional osteological literature and at scientific congresses, however, to many physicians the substance of these terms remains unclear. The aim of this article is to explain in more detail the biomechanical methods used in determining bone quality as well as to describe basic characteristic bone qualities resulting in the application of these methods.

Biomaterials may be divided into brittle and tenacious ones according to their mechanical reaction to simple tensile and compression stress. Brittle materials deteriorate by very slight and permanent deformations. Tenacious materials deteriorate after a major and permanent deformation. Mechanical properties of biomaterials are characterised by stress – strain curves, produced on the basis of testing the biomaterials by progressive tensile and compressive load. According to these curves, the bone counts among brittle materials. Stress – strain curves under tensile and compressive load for brittle materials comprise proportional limit, elastic limit, and breaking strength. Regarding the definitions of elasticity and strength, we can state that if the total strain, produced by external load, exceeds the elastic limit of the strained bone, the bone suffers plastic deformation, and exceeding the strength limit leads to bone fracture.

To understand completely the mechanical properties of biomaterials during tensile and compressive load, we have to put them to tests.

The quality of healthy or osteoporotic bone can be *objectively determined* in biomechanics by destructive methods.

Test bone or its sample is inserted into a special device so that two opposite ends are firmly fixed in the direction of linear axis. Afterwards, the tested sample is gradually loaded by growing tensile or compressive forces until these forces reach the values by which the bone breaks. Measurement results are illustrated by diagrams.

DETERMINATION OF MECHANICAL PROPERTIES OF BONES

Diagrams are constructed in two variants according to the character of variables put onto coordinate axes *x* and *y*:

1. Diagram of tension (compression) test
2. Stress – strain curve under tensile (compressive) loading

Diagram of Compression Test for Bone (Graph 1a) [1-3]

We put the gradually increasing values of *compressive force* F_c on the bone in Newtons (N) to the axis y, and the values of the *absolute shortening of bone* Δl (mm) to the axis x.

Absolute shortening:

$$\Delta l = \frac{F \times l}{E \times P}$$

Up to the point of yielding (A), the shortening of the bone is directly proportional to the compressive force magnitude, and whenever the application of compressive force quits, the bone regains its original length, as its deformation has been only temporary (reversible). This part of the diagram is called: bone elasticity. Beyond the point of yielding there is a plastic region, corresponding to bone toughness. Increasing the compressive force results in further shortening of the bone, however, permanent deformations caused by the compressive force persist even after the removal of the compressive force. It is a case of plastic deformation of the bone. The bone begins to crack in the breaking point (B), and increasing the magnitude of applied compressive force completely disrupts its integrity.

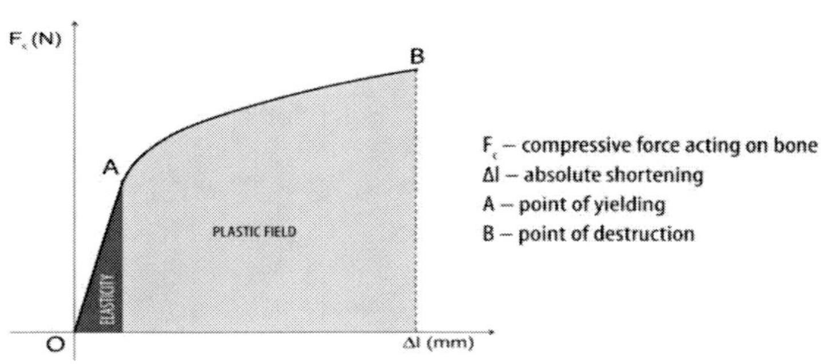

Graph 1a. Diagram of compression test for bone.

Stress – Strain Curve under Compressive Loading for Bone (Graph 1b) [4-7]

The stress values ***σ*** (*Ncm*$^{-2}$) developed during loading the bone by compressive force are put on the axis ***y***, and the values of relative shortening ***ε*** (non-dimensional number) on the axis *x*.

Stress:

$$\sigma = \frac{F}{P}$$

Relative shortening:

$$\varepsilon = \frac{\Delta l}{l}$$

Hooke's law (law of elasticity):

$$\Delta l = \frac{F \times l}{E \times P}$$

Elasticity modulus under compression:

$$E = \frac{F \times l}{\Delta l \times P} \quad E = \frac{\sigma}{\varepsilon}$$

F – compression force applied to the bone
P – the section area of the bone, to which the force is applied
l – the length of the test bone sample
E – proportionality coefficient (elasticity modulus under compression)

The diagram includes all magnitudes of tension, characterising mechanical properties of the bone: *proportional limit, elastic limit, breaking strength.*

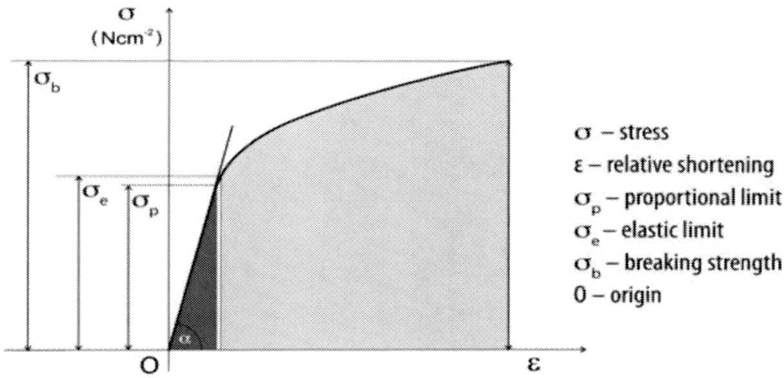

Graph 1b. Stress – strain curve under compressive loading for bone.

The highest value of stress in the bone conditioned by its compressive load, when the Hooke's law still applies, is called a proportional limit. The highest value of stress, when still elastic reversible deformations occur, is called an elastic limit. Points for proportional limit and elastic limit are so close that they are usually considered to be identical.

The stress generated by compressive force, which causes the development of first cracks in the bone, is called a strength limit.

ELASTICITY MODULUS (E) UNDER COMPRESSION FOR BONE

The ratio of the stress under compression for bone (σ) and the relative shortening of bone (ε) is defined as a modulus of elasticity under compression for bone and is a constant of the Hooke's law.

Elasticity modulus is proportional to the gradient (inclination angle tangent α) of the linear part of the diagram curve from zero stress (σ_0) up to proportional limit (σ_p).

Elasticity modulus under compression for bone is a number characterising the bone behaviour under compressive load. The higher the elasticity modulus, the higher stress is needed to achieve bone deformation.

DIAGRAM OF TENSION TEST FOR BONE

(GRAPH 2A) [1- 4]

Gradually increasing values of *tensile force* F_t on the bone in Newtons (N) are put onto the axis ***y***, and the values of the *absolute extension of bone* Δl (mm) on the axis x.

Absolute extension (Δl):

$$\Delta l = \frac{F \times l}{E \times P}$$

Up to the point of yielding (A), the extension of the bone is directly proportional to the tensile force magnitude, and whenever the application of tensile force quits, the bone regains its original length, as its deformation has been only temporary. This part of the diagram is called: bone elasticity. Beyond the point of yielding there is a plastic region, corresponding to bone toughness. Increasing the tensile force results in further extension of the bone, however, permanent deformations caused by the tensile force persist even after the removal of the tensile force. It is a case of plastic deformation of the bone. The bone begins to crack in the breaking point (B), and increasing the magnitude of applied tensile force completely disrupts its integrity.

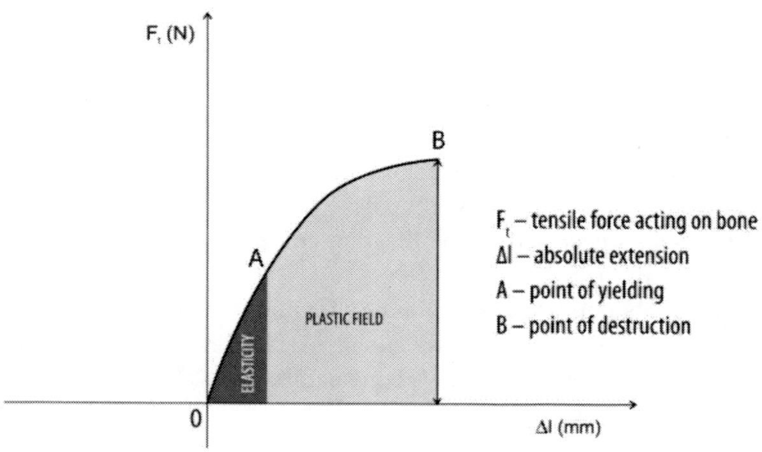

Graph 2a. Diagram of tension test for bone.

Stress – Strain Curve under Tensile Loading For Bone (Graph 2b) [7-9]

The stress values σ (Ncm^{-2}) developed during loading the bone by tensile force are put on the axis ***y***, and the values of relative extension ε (non-dimensional number) ε on the axis ***x***.

Tension:

$$\sigma = \frac{F}{P}$$

Relative extension:

$$\varepsilon = \frac{\Delta l}{l}$$

Hooke's law (law of elasticity):

$$\Delta l = \frac{F \times l}{E \times P}$$

Elasticity modulus under tension:

$$E = \frac{F \times l}{\Delta l \times P} \quad E = \frac{\sigma}{\varepsilon}$$

F – tensile force applied to the bone
P – the section area of the bone, to which the force is applied
l – the length of the test bone sample
E – proportionality coefficient (elasticity modulus under tension)

The diagram includes all magnitudes of stress, characterising mechanical properties of the bone: *proportional limit, elastic limit, breaking strength.*

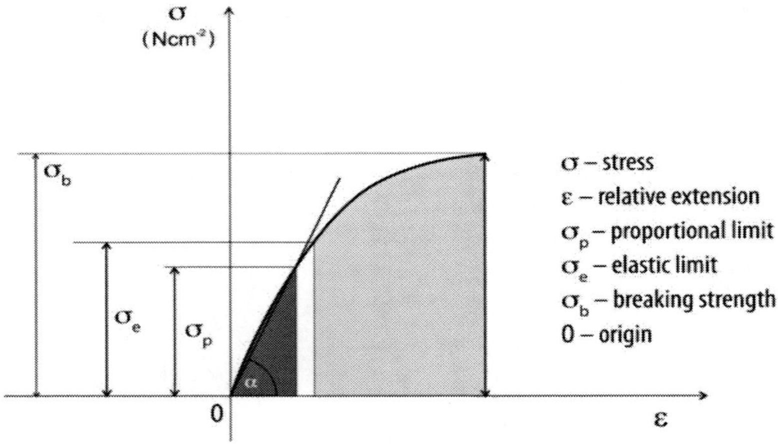

Graph 2b. Stress – strain curve under tensile loading for bone.

The highest value of stress in the bone conditioned by its tensile load, when the Hooke's law still applies, is called a proportional limit. The highest value of stress, when elastic reversible deformations still occur, is called an elastic limit. Points for proportional limit and elastic limit are so close that they are usually considered to be identical.

The stress generated by tensile force, which causes the development of first cracks in the bone, is called a strength limit.

Elasticity Modulus (E) under Tension for Bone

The ratio of the stress under tensile force for bone (σ) and the relative extension of bone (ε) is defined as a modulus of elasticity under tension for bone and is a constant of the Hooke's law.

Elasticity modulus is proportional to the gradient (inclination angle tangent α) of the linear part of the diagram curve from zero stress (σ_0) up to proportional limit (σ_p).

Elasticity modulus under tension for bone is a number characterising the bone behaviour under tensile load. The higher the elasticity modulus, the higher stress is needed to achieve bone deformation.

When comparing the diagram of stress and strain under compression and tension for bone (the diagram of compression and tension tests for bone), we find out that the bone has higher resistance to compressive than to tensile load.

Conclusion

The characteristics of osteoporotic bone quality, which determines risk of fractures, is based on biomechanical principles.

Recognition of biomechanical characteristics of bone quality (proportional limit, elastic limit, breaking strength) helps physicians to understand the importance of measuring the new parameters in bone densitometry by DXA (dual energy X-ray absorptiometry) such as: femur strength index (FSI), and safety factor (SF). Formulas of these parameters use the characteristics from stress – strain curves for loading the bone by tension and compression [10, 11]. More accurate estimate of the risk of osteporotic fractures development is helped by FSI and SF, as compared to currently used BMD (bone mineral density) values, and these parameters are applied in osteological practice with success.

On the contrary, misinterpretation of basic biomechanical relations by medical professional community leads often to the lack of interest in the new methods based on biomechanical principles and delays their implementation in clinical praxis. Even clinical physicians have to understand that biomechanics is becoming an indivisible part of medicine and is contributing to the improvement of diagnostic and therapeutic methods.

References

[1] Cook RB, Zioupos P: The fracture toughness of cancellous bone. *J. Biomech.* 2009;42(13):2054-60.

[2] Barak MM, Currey JD, Weiner S : Are tensile and compressive Young's moduli of compact bone different? *J. Mech. Behav. Biomed. Mater.* 2009;2(1):51-60.

[3] Bono CM, Einhorn TA : Overview of osteoporosis: pathophysiology and determinants of bone strength. *(Review) Europ. Spine J.* 2003; 12: 90-96.

[4] Chao EY, Inoue N, Koo TK, Kim YH: Biomechanical considerations of fracture treatment and bone quality maintenance in elderly patients and patients with osteoporosis. *Clin. Orthop. Rel. Res.* 2004; 25: 12-25.

[5] Lohfeld S, Barron V, Mc Hugh PE: Biomodels of bone: A review. *An. of Biomed. Engineering* 2005; 10: 145–150.

[6] Burr DB: Biomechanics. *Clin. Reviews in Bone and Miner. Metab.* 2006; 3:112–117.

[7] Wang JF. Modelling Young's modulus for porous bones with microstructural variation and anisotropy. *J. Mater. Sci. Mater. Med.* 2010;21(2):463-72.

[8] Özkaya N, Nordin M: *Fundamentals of biomechanics*: Equilibrium, motion and deformation, 2nd ed, Springer – Verlag, New York, 1999, 45 – 51.

[9] Routh RH, Rumancik S, Pathal RD, Burshell LA, Nauman AE: The relationship between bone mineral density and biomechanics in patients with osteoporosis and scoliosis *Osteoporos. Int.* 2005;16(12):116-122.

[10] Faulkner KG, Wacker WK, Barden HS, Simonelli C, Burke PK, Ragi S, Del Rio L: Lemur strength index predicts hip fracture independent of bone density and hip axis length. *Osteoporos. Int.* 2006;17(4):593–599.

[11] Bevill G, Farhamand F, Keaveny TM: Heterogeneity of yield strain in low-density versus high-density human trabecular bone. *J. Biomech.* 2009;42(13):2165-70.

Chapter IX

Statistical Tests for Estimating Osteoporotic Fracture Risk *(Survey For Physicians)*

Abstract

The article presents an analysis of statistical methods used for estimating fracture risk in patients with osteoporosis. Mathematical relations of different methods are explained (risk – R, risk ratio - RR, RD – risk difference, odds - O, odds ratio - OR, Yule's Q, Yule's Y, logistic model). What is important to keep in mind is that: 1) relative risk and odds ratio are statistics that only describe an association, not causation; 2) relative risk and odds ratio refer to a population, not to an individual patient; 3) the studies of small groups are more likely to find an association that might actually just be due to chance, the larger the groups, the less likely the association between a risk factor and an outcome (fracture); 4) when the incidence of an outcome of interest in the study population is low (< 10%), the OR is close to the RR, the more frequent the outcome becomes, the more the OR will overestimate the RR when it is more than 1 or underestimate the RR when it is less than 1. Sophisticated statistical packages are available which can calculate many of the tests of association but the problem is that the investigator must know which the desirable is. The incorrect option of statistical analysis, the incorrect interpretation of risk ratio or odds ratio and overestimation of the importance of a risk factor may lead to unintentional errors in the economic analysis of potential programs or treatments in osteoporosis. This article could be a contribution for investigators, who are concerned with assessment of fracture risk.

Keywords: relative risk of fracture – absolute risk of fracture- odds ratio –risk diference- Yule's Q - Yule's Y – osteoporosis

INTRODUCTION

There increase the costs of the complex care of patients with osteoporotic fractures from year to year. The detection and adequate treatment of patients with a higher fracture risk is of great interest. The criteria for the evaluation of a fracture risk have constantly been reassessed and updated with one goal – to maximise the validity and accuracy of the applied methods.

The aim of the article is to present the statistical methods enabling the analysis of fracture risk estimation in patients with osteoporosis in an investigated population. Mathematical relations of different methods are explained. At present there are two priorities in the centre of interest in the research of osteoporotic patients:

1. observation of the influence of risk factors on the bone mineral density (BMD) values and on the onset of fractures (observational studies)
2. observation of the influence of antiosteoresorptive or bone formation stimulating drugs on BMD values and the reduction of fracture risk (interventional studies – trials)

STATISTICAL ANALYSIS

The evaluation of the fracture risk for patients with osteoporosis can be realized in both observational and interventional studies by retrospective (case – control) or prospective (cohort, incidence) studies. A retrospective study determines: How did the exposure to risk factors influence the fracture? How did the antiosteoresorptive or bone formation stimulating drugs decrease the incidence of fractures? A prospective study determines: Will the exposure to risk factors condition the incidence of a fracture in the future? Will the antiosteoresorptive or bone formation stimulating drugs prevent the patient from the fracture in the future? [1-3].

It is preferred to use the odds ratio in retrospective studies and the relative risk or relative risk corrected to odds ratio in prospective studies (see eq. 2, 3 and 7). In the observation of the influence of fracture risk factors, or the influence of drugs on the reduction of fracture incidence we are working with dichotomous variables. Dichotomous variables characterise the quality of the observed parameter and we classify them by "yes" or "no", and in the statistical evaluation by the values of 0 and 1.

Association among dichotomous variables can be judged by contingency tables 2x2 and can be measured by several statistical methods:

- risk (R)
- risk ratio (RR)
- odds (O)
- odds ratio (OR)
- approximation (correction) of OR to RR (RR_{cor})
- Yule's Q and Yule's Y ratio (Q, Y)

As an illustration we present a simple 2x2 contingency table with dichotomous variables in the article (Table 1).

- risk factor – present, absent (+, -)
- fracture – present, absent (+, -)

Table 1. Contingency table 2x2

Risk factor	*Fracture*	
	+	–
+	a	**b**
–	c	**d**

a, b, c, d – frequencies.

Risk (R) [4-6]

Risk R_1 -- yields the risk for the fracture in patients with the risk factor.
Risk R_2 -- yields the risk for the fracture in patients without the risk factor.

$$R_1 = \frac{a}{a+b} \qquad \text{(eq. 1a)}$$

$$R_2 = \frac{c}{c+d} \quad \text{(eq. 1b)}$$

$$R = probability(p) = \frac{x}{n} \quad \text{(eq. 1c)}$$

$$R(\%) = probability(p) = \frac{x}{n} \cdot 100\% \quad \text{(eq 1d)}$$

$$R\langle 0,1 \rangle$$

The risk R_1 is a ratio of the number of patients with fractures exposed to risk factor (*a*) to the total number of observed patients exposed to risk factor (*a* + *b*). The risk R_2 is a ratio of the number of patients with fractures without exposure to risk factor (*c*) to the total number of patients not exposed to risk factor (*c* + *d*) (Tab. 1). From the statistical point of view it is actually the risk expressed as a probability (*p*) – the number of positive or negative cases (*x*) (event rates) to the total number of observed cases (*n*).

Risk values vary within the interval $\langle 0,1 \rangle$

Relative Risk = Risk Ratio (RR) [4, 5, 6, 7, 8, 9]

RR – is the ratio of the risk for the fracture in patients with the risk factor, to the risk for the fracture in patients without the risk factor. In the equation 2a, the RR is expressed also by means of the probabilities ratio.

$$RR = \frac{R_1}{R_2} = \frac{\frac{a}{a+b}}{\frac{c}{c+d}} \quad \text{(eq. 2)}$$

$$RR = \frac{p_1}{p_2} = \frac{\frac{x_1}{n_1}}{\frac{x_2}{n_2}} \quad \text{(eq. 2a)}$$

The RR takes on values between zero (0) and infinity (∞). One (RR = 1) is the neutral value and means that there is no difference between the groups compared, close to zero (RR $\approx$ 0) or infinity (RR $\approx$ ∞) means a large difference between the two groups on the variable concerned. A risk ratio larger than one (RR > 1) means that group one has a larger proportion than group two, if the opposite is true the risk ratio will be smaller than one (RR < 1). The RR multiplicated by 100% gives the percentage proportion in classification between one and group two. It needs to be calculated for each value of RR the 95% confidence interval (C.I.).

Risk Difference (RD)

$$RD = R_1 - R_2 \qquad \text{(eq. 2b)}$$

RD gives difference between risk R_1 for the fracture in patients with the risk factor, to the risk R_2 for the fracture in patients without the risk factor.

Example No 1

We can compare two groups with different risks (R_1, R_2) for hip fracture from data in prospective epidemiological study. For example we have investigated that, the risk (R_1) for patients suffering from hip fracture with BMD T – score = -1. 4 SD is 0. 05 (5 %) (group A), while the risk (R_2) in the group B with BMD T – score = - 2,6 SD is 0.20 (20%). The RR equals 0, 25 ($R1/R_2$ = 0.05 / 0.20).

The difference from neutral value 1 is 0. 75 (1 – 0. 25 = 0. 75 .100 = 75%).

75% fewer patients in group A are in the relative risk to suffer hip fracture.

We chose for T - score in the example No 1 only the categorical level without the C.I. for simple interpretation.

Odds (O) [8 -14]

O – is given as a probability (p) that the fracture will happen, divided by the probability that the fracture will not happen.

$$O = \frac{p}{1-p} \quad \text{(eq. 3)}$$

In the eq. 3 the probability (p) can be substituted by the risk (R)

$$O = \frac{R}{1-R} \quad \text{(eq. 4)}$$

Interpretation of Odds Results

Odds values (O) vary within the interval $\langle 0, \infty \,)$. If the value O > 1, there are higher odds that the fracture will occur in the observed group. If the value O < 1, there are higher odds that the fracture will not occur. The odds that the fracture will definitely occur equal infinity ($O = \infty$), and the odds that the fracture will not occur at all equal zero (O = 0).

The ratio of two odds values is expressed as odds ratio.

Odds Ratio (OR) [8-14]

OR – is the ratio of the odds of the fracture for patients with the risk factor, to the odds of the fracture for patients without the risk factor.

$$OR = \frac{O_1}{O_2} = \frac{\dfrac{p_1}{1-p_1}}{\dfrac{p_2}{1-p_2}} = \frac{\dfrac{\frac{a}{a+b}}{1-\left(\frac{a}{a+b}\right)}}{\dfrac{\frac{c}{c+d}}{1-\left(\frac{c}{c+d}\right)}} = \frac{\frac{a}{b}}{\frac{c}{d}} = \frac{a.d}{b.c} \quad \text{(eq. 5)}$$

The OR takes values between zero (0) and infinity (∞). One (OR = 1) is the neutral value and means that there is no difference between the groups compared, close to zero ($OR \approx 0$) or infinity ($OR \approx \infty$) means a large difference. An OR larger than one (OR > 1) means that group one has a larger proportion than group two (positive association), if the opposite is true the OR will be smaller than one (OR < 1) (negative association).

Example No 2

If 10 % of patients in group A1 having BMD T- score (-1. 4 SD) suffer from hip fracture, than the odds (O_1) for suffering from hip fracture is 0.11 (0.10 / 0.90) (eq. 3). The patients in group A1 have 0.11 times higher chance of getting fracture than of not getting fracture.

If 15% of patients in group B1 with BMD T – score (- 2. 6 SD) suffer from hip fracture, than the odds (O_2) for suffering from hip fracture is 0.176 (0.15 / 0.85) (eq. 3).

The patients in group B1 have 0.176 times higher chance of getting fracture than of not getting fracture. The odds ratio (OR) equals 0.625 ($O_1/O_2 =$ 0.11 / 0.176). The patients in group A1 have 0.625 times higher chance of getting fracture than in group B1. The inverse of the odds ratio ($O_2/O_1 = 1/OR$) equals 1.6 (0.176 / 0.11). The patients in group B1 have 1.6 times higher chance of getting fracture than patients in group A1.

OR indicates how much the odds of the fracture for patients with the risk factor is higher than the odds of the fracture for patients without the risk factor.

$$OR = \frac{\frac{a}{b}}{\frac{c}{d}} = \frac{\frac{a}{c}}{\frac{b}{d}} = \frac{\frac{d}{b}}{\frac{c}{a}} = \frac{\frac{d}{c}}{\frac{b}{a}} = \frac{a.d}{b.c} \qquad \text{(eq. 6)}$$

In equation (eq. 6) we see that the OR can be interpreted in various ways. We can compare the patient groups *a, b, c, d*, in the contingency table (Table 1) in a horizontal or vertical direction, but not in cross direction, regardless of whether one is concerned with risk or without risk, or whether one has a fracture or has not – the outcome is the same. If an observation is higher in value as another observation on both the horizontal and the vertical marginals, the pair of observations is called concordant; if this is not the case the pair is discordant.

Some research papers present an incorrect opinion that OR and RR are identical statistical methods [15]. The eq. 2 and eq. 5 in our analysis indicate the opposite. Relative risk is not the same as odds ratio.

The graph 1 shows that when the incidence of an outcome of interest in the study population is low (< 10%), the OR is close to the RR. The more frequent the outcome becomes, the more the OR will overestimate the RR when it is more than 1 or underestimate the RR when it is less than 1.

The RR is easier to interpret than the OR. However, in practice the odds ratio is used more often. This has to do with the fact that the OR is more closely related to frequently used statistical techniques such as logistic regression. Also, the OR has the attractive property that, however is turned the table (Table 1), it will always take on the same value (eq. 6).

There must be calculated the 95% confidence interval (C.I.) for each value of OR and RR (for example in SAS^{R} system).

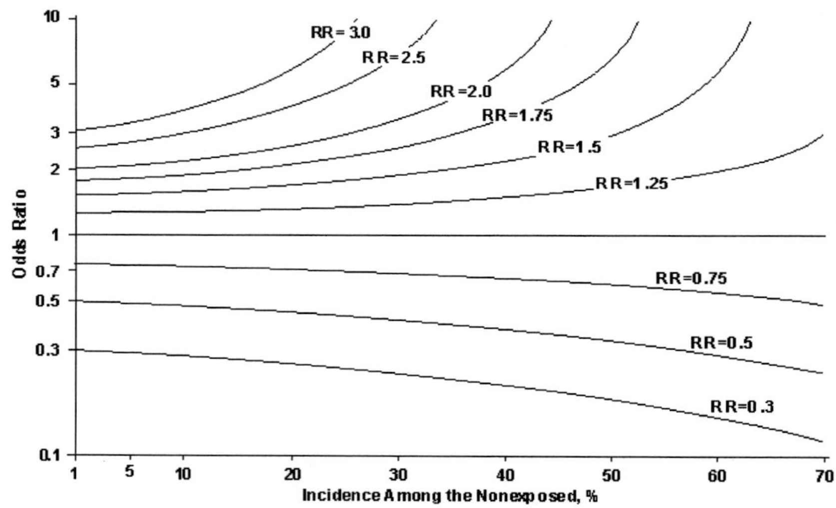

Graph 1. The relationship between risk ratio (RR) and odds ratio (OR) by variable incidence of the outcome. (Zhang J, Kai F : What is the relative risk? JAMA, 1998, 280, 19, 1690 – 91).

WHEN TO USE RR AND OR IN THE EVALUATION OF THE ASSOCIATION OF DICHOTOMOUS VARIABLES?

(TABLE 2)

The relevant literature states that OR is more appropriate for the evaluation of retrospective studies and RR or RR_{cor} (eq. 7) is better for prospective studies [13-16].

Corrected RR (RR_{cor}) [5]

In a cohort study, if the incidence of outcome is more than 10% (Graph 1) and the OR is more than 2.5 or less than 0.5, a correction of the OR may be desirable to more appropriately interpret the magnitude of an association.

$$RR_{cor} = \frac{OR}{(1 - p_0) + (p_0 \cdot OR)} \quad \text{(eq. 7)}$$

p_0 – indicates the incidence of the outcome of interest in the not exposed group, it is the ratio of patients without a risk in a group to the total in the group.

Yule's Q, Yule's Y (Q, Y) [13]

If we want to compare the influence of two risk factors separately on the fracture onset as a comparison of two OR (two contingency tables: 2 x 2 and 2 x 2), we have to evaluate which risk factor has higher OR to condition the fracture onset (which drug has higher OR to reduce the risk of a fracture) by using two mathematical and statistical methods – Yule's Q and Yule's Y.

$$Q = \frac{OR - 1}{OR + 1} \quad \text{(eq. 8)}$$

$$Y = \frac{\sqrt{OR} - 1}{\sqrt{OR} + 1} \quad \text{(eq. 9)}$$

Both methods compare two OR. The Q and Y transform the odds to the range $\langle +1, -1 \rangle$ with Q (Y) = 0 when OR = 1, Q (Y) = - 1 when OR = 0 and Q (Y) approaches 1 as OR approaches infinity, Q and Y are the gamma statistic.

Yule's Q and Y are based on the OR and a symmetric measure taking on values between -1, and +1. One (1) implies perfect negative or positive association, zero (0) no association. By Yule's Y the measure tends to estimate association more conservatively than Yule's Q and has little substantive or theoretical meaning.

Transformation of the result of the ration of two OR from the original interval $\langle 0, \infty \,)$ into the interval $\langle +1, -1 \rangle$ makes its interpretation more clear.

Table 2. Types of studies frequently used to determine the fracture risk and recommended statistical analysis

Studies	*Recommended statistical analysis*
case – control studies (retrospective)	OR Yule's Q, Yule's Y
cohort studies (prospective, incidence)	RR, RR_{cor},

More Than Two Risk Factors

Logistic model [8-16]

When observing k risk factors (R_1, R_2, ...R_k) and their influence upon the fracture onset at once, we can use k contingency tables (type n_i x 2), where n_i is the number of levels of risk factors R_i (i = 1, 2, 3, ...k).

For more than two risk factors these possibilities are to calculate:

- OR for all pairs separately in each group one risk factor (present, absent) and one dichotomous outcome (fracture, no fracture) → simple analysis
- logistic regression analysis

Logistic regression uses the explanatory variables, which can be categorical or continuous, to predict the probability that the outcome or response variable takes on a given value. In other words, logistic regression is designed to describe probabilities associated with the values of the outcome variable. Probabilities are bounded by 0 and 1, so a linear model cannot be used because linear functions are inherently unbounded. The solution to this problem is to transform the probabilities to *logits,* which are unbounded, so that a linear regression model can be used because the logits are linear in the parameters.

The logistic regression model describes the relationship between a categorical outcome (response) variable Y and a set of explanatory variables (X_i). It models the conditional probability of a successful outcome, which is denoted $p = \mathrm{P}(Y = 1/X = x)$. The relationship between probability p and a set of explanatory variables (X_i) is not linear

$$p = \frac{\exp(\beta_0 + \beta_1 X_1 + \beta_2 X_2 + \ldots + \beta_k X_k)}{1 + \exp(\beta_0 + \beta_1 X_1 + \beta_2 X_2 + \ldots + \beta_k X_k)}$$

The relationship between *logit* and a set of explanatory variables (X_i) is linear:

$$Logit = \ln(odds) = \ln \frac{p}{1-p} = \beta_0 + \beta_1 X_1 + \beta_2 X_2 + \ldots + \beta_k X_k$$

Unlike linear regression, the categorical response is not normally distributed and the variances are not the same. Also, logistic regression usually requires a more complex iterative estimation method called maximum likelihood to estimate the parameters than linear regression. This method finds the parameter estimates that are most likely to occur given the data. It accomplishes this by maximizing the likelihood function that expresses the probability of the observed data as a function of the unknown parameters.

Individual risk of fracture

Individual risk of the fracture in a patient has been determined in the last few years through a popular form of various questionnaires, where, besides the bone mineral density values in the total hip area, the points are given for the answers concerning the presence of a risk factor in the anamnesis – e.g., age, weight, a fracture sustained after the age of 50, hip fracture in mother after the age of 50, biochemical markers of osteoresorption, smoking, and the like [17]. Each questionnaire after its completion has to be tested for sensitivity, specificity, positive predictive value, and negative predictive value – the most proper statistical test methods include the $\chi 2$ - square test and the ADALINE programme [18, 19].

CONCLUSION

In this article we have reviewed the methods used to analyse associations between variables – risk factor and fracture in 2 x 2 contingency tables. What is important to keep in mind is that:

1. relative risk and odds ratio are statistics that only describe an association, not causation
2. relative risk and odds ratio refer to a population, not to an individual patient
3. when the incidence of an outcome of interest in the study population is low (< 10%), the OR is close to the RR, the more frequent the outcome becomes, the more the OR will overestimate the RR when it is more than 1 or underestimate the RR when it is less than 1.

It is important that before starting a research, its annotation should be seriously prepared with a clear intention and selection of a proper method for statistical evaluation. Sophisticated statistical packages are available which can calculate many of the tests of association but problem is that the investigator must know which is the desirable one. The incorrect option of statistical analysis, the incorrect interpretation of risk ratio or odds ratio and overestimation of the importance of risk factor may lead to unintentional errors in the economic analysis of potential intervention programmes or treatments in osteoporosis.

REFERENCES

[1] Kanis JA, Johnell O, Oden A, Jonsson B, Dawson A, Dere W : Risk of hip fracture derived from relative risk: An analysis applied to the population of Sweden. *Osteoporos. Int.* 2000;11: 120–127.

[2] Kanis JA, Johnell O, Oden A, Oglesby AK, De Laet C, Jonsson B: Ten year probabilities of osteoporotic fracture according to BMD and diagnostic thresholds, *J. Bone Miner. Res.* 2001; 16 (Suppl 1): 194.

[3] Mc Nutt LA, Wu CH, Xue X, Hafner JP: Estimating the relative risk in cohort studies and clinical trials of common outcomes. *Am. J. Epidemiol.* 2003;157: 940-943.

[4] Sinclair JC, Bracken MB: Clinically useful measures of treatment effect in binary analyses of randomized trials. *J. Clin. Epidemiol.* 1994; 47:881–889.

[5] Zhang J, Yu KF: What is relative risk? A method of correcting the odds ratio in cohort studies of common outcomes. *JAMA* 1998; 280:1690–1691.

[6] Greenland S: Model – based estimation of relative risks and other epidemiologic measures in studies of common outcomes and in case – control studies. *Am. J. Epidemiol.* 2004;160(4):301–305.

[7] Melton JL III, Atkinson E, O'Connor MK, O'Fallon WM, Riggs BL: Bone density and fracture risk in men. *J. Bone Miner. Res.* 1998;13:1915-1917.

[8] Hilgers RD, Bauer P, Scheber V: Einfuhrung in die medizinische Statistik Springer Verlag Berlin Heidelberg, 1. *Ausgabe*, 2003, 217 – 220.

[9] Kleinbaum DG, Klein M : *Logistic regression*. Statistics for Biology and Health, Springer, 2nd edition 2002, corr. 5th printing 2009;2 – 45.

[10] Lee J: Odds ratio or relative risk for cross – sectional data? *Int. J. Epidemiol.* 1994; 23: 201–203.

[11] Bland JM, Altman DG: The odss ratio. *Brit. Med. J.* 2000;320:1468-1470.

[12] Davies HTO, Crombie IK, Tavakoli M: When can odds ratios mislead? *BMJ* 1998; 316:989–991.

[13] Last JM: *A dictionary of epidemiology*, 4 th ed, RA Spasoff, SS Harris and MC Thuriaux eds. Oxford, Oxford University Press, 2001, 45 – 82.

[14] Jaeschke, Guyatt G, Shannon H, Walter S, Cook D, Heddle N: Assessing the effects of treatment: measures of association. *CMAJ* 1995; 152:351–357.

[15] Mikulecky M: Association of nominal variables – contingency tables. *Euro Rehab*. 2003; 9:88-96.

[16] Egger M, Davey Smith G, Philips AN: Meta analyses principles and procedures. *BMJ* 1997; 315:1533–1537.

[17] Black DM, Steinbuch L, Palermo L, Dargent – Molina P, Lindsay R, Hoseyni MS, Johnell O: An assessment tool for predicting fracture risk in postmenopausal women. *Osteoporos. Int.* 2001;12:519–528.

[18] Widrow B, Smith FW: *Pattern recognizing control systems*. In: Tou JT., Wilcox RH. Computer and Information science. Collected papers on learning adaptation and control in information systems. 2nd ed, Washington, Sparten Books, 1964, 40 – 47.

[19] Schumacher M, Schulgen G: *Methodik klinischer Studien*. Springer Berlin,2 nd ed. 2006, 71-104.

Index

D

E

F

N

O

P

Q

R

S

T

U

V

W

Y